STUDENT SOLUTIONS MANUAL

FIFTH EDITION

CALCULUS

WITH ANALYTIC GEOMETRY

STUDENT SOLUTIONS MANUAL

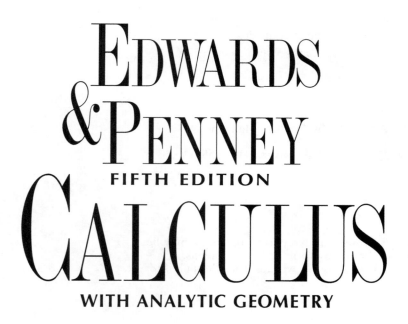

EDWARDS & PENNEY

FIFTH EDITION

CALCULUS

WITH ANALYTIC GEOMETRY

PRENTICE HALL, UPPER SADDLE RIVER, NJ 07458

Executive Editor: George Lobell
Production Editor: Barbara Kraemer
Supplement Cover Designer: Marianne Frasco
Special Projects Manager: Barbara A. Murray
Supplement Cover Manager: Paul Gourhan
Manufacturing Buyer: Alan Fischer
Assistant Editor: Audra Walsh

Printed in the United States of America

10 9 8 7 6 5

ISBN 0-13-757774-5

Prentice-Hall International (UK) Limited, *London*
Prentice-Hall of Australia Pty. Limited, *Sydney*
Prentice-Hall Canada, Inc., *London*
Prentice-Hall Hispanoamericana, S.A., *Mexico*
Prentice-Hall of India Private Limited, *New Delhi*
Prentice-Hall of Japan, Inc., *Tokyo*
Pearson Education Asia, *Singapore*
Editora Prentice-Hall do Brasil, Ltda., *Rio de Janeiro*

Contents

Preface

This manual contains solutions to problems numbered 1, 4, 7, 10, ..., in Chapters 1 through 15 of *Calculus with Analytic Geometry*, 5th edition (1998), by C. Henry Edwards and David E. Penney. Because the answers to most odd-numbered problems are given in the *Answer* section of the text, odd-numbered solutions are omitted from this manual whenever the answer alone to such a problem is sufficient. When a solution is long but not difficult, only the outline of a solution may appear. When a suggestion alone is sufficient, we often give only a suggestion, or in a few cases an alternative solution. Rarely, if an answer would spoil the problem for the solver, we omit it.

Many calculus problems can be solved by more than one method. We generally use the most natural method, but in rare instances offer a "clever" method in its place, but only when its educational value justifies such a substitution. In particular, in working these problems we used only those techniques developed in earlier sections (with rare exceptions, and then only when the alternative method is accessible and useful to the reader).

We gratefully acknowledge the encouragement, advice, and assistance of our colleagues and students in helping us to correct errors in previous editions of this manual. Special thanks go to Nancy and Mary Toscano of Toscano Document Engineers (Somerville, MA), whose staff checked the worked-out examples, answers to odd-numbered problems, and solutions to most of the problems in this manual. We also checked most solutions with *Mathematica*. Will Kazez and Ted Shifrin helped us with TEX and Adobe Illustrator, the two programs used to prepare the final version of this manual. Most of all, we thank Carol W. Penney for typesetting the text of all solutions manuals.

C. Henry Edwards (hedwards@math.uga.edu)

David E. Penney (dpenney@math.uga.edu)

The University of Georgia, Athens

August 1997

Typeset by $\mathcal{AMS}$-TEX

Chapter 1: Functions and Graphs
Section 1.1

1. (a) $f(-a) = -1/a$

    (b) $f(a^{-1}) = a$

    (c) $f(\sqrt{a}) = 1/\sqrt{a}$

    (d) $f(a^2) = 1/(a^2)$

4. (a) $f(-a) = \sqrt{1 + a^2 + a^4}$

    (b) $f(a^{-1}) = \sqrt{1 + a^{-2} + a^{-4}}$

    (c) $f(\sqrt{a}) = \sqrt{1 + a + a^2}$

    (d) $f(a^2) = \sqrt{1 + a^4 + a^8}$

7. $\sqrt{a^2 + 16} = 5$; $a^2 = 9$; $a = \pm 3$

10. $2a^2 - a + 4 = 5$; $2a^2 - a - 1 = 0$; $(2a + 1)(a - 1) = 0$; $a = 1$ or $a = -\frac{1}{2}$

13. $f(a + h) - f(a) = (a + h)^2 - a^2 = 2ah + h^2$

16. $f(a + h) - f(a) = \dfrac{2}{a + h + 1} - \dfrac{2}{a + 1} = -\dfrac{2h}{(a + h + 1)(a + 1)}$

19. Because the exponent takes on all integral values, odd and even, the range of f is the two-element set $\{-1, 1\}$.

22. The set R of all real numbers

25. The domain of f consists of those numbers x for which $3x - 5 \geq 0$: The interval $\left[\frac{5}{3}, \infty\right)$.

28. The set of all real numbers other than -2; in interval notation, $(-\infty, -2) \cup (-2, \infty)$.

31. The set R of all real numbers, because $x^2 + 9 \geq 0$ for all real x.

34. We require $\dfrac{x + 1}{x - 1} \geq 0$ and that $x \neq 1$. The former condition holds when numerator and denominator have the same sign, which implies $x > 1$ or $x \leq -1$. Thus the domain of f consists of those numbers x such that $x \leq -1$ together with those numbers x such that $x > 1$.

37. The radius of the circle is $r = \sqrt{A/\pi}$ and its circumference is $C = 2\pi r$. Therefore
$$C(A) = 2\sqrt{\pi A}, \qquad A \geq 0 \quad (\text{or } A > 0).$$

40. If the rectangle has base x and height y, then $2x + 2y = 100$, so $y = 50 - x$.

43. Denote the height of the box by y and note that $x^2 y = 324$. The cost C of the box is the sum of the cost $(2x^2)$ of its base, four times the cost (xy) of one of its sides, and the cost (x^2) of its top. Thus $C = 3x^2 + 4xy$, and so
$$C(x) = 3x^2 + \frac{1296}{x}, \qquad x > 0.$$

46. If the box has height y, then $2x^2 + 4xy = 600$. Because the volume V is $V = x^2 y$, it follows that
$$V(x) = \tfrac{1}{2}x\left(300 - x^2\right), \qquad 0 < x \leq \sqrt{300} \quad \left(\text{or } 0 < x < \sqrt{300}\right).$$

49. Recall that the total daily production of the oil field is $p(x) = (20 + x)(200 - 5x)$ if x new wells are drilled (where x is an integer satisfying $0 \leq x \leq 40$). Here is a table of *all* the values of the function p:

x	0	1	2	3	4	5	6	7
p	4000	4095	4180	4255	4320	4375	4420	4455

x	8	9	10	11	12	13	14	15
p	4480	4495	4500	4495	4480	4455	4420	4375

x	16	17	18	19	20	21	22	23
p	4320	4255	4180	4095	4000	3895	3780	3655

x	24	25	26	27	28	29	30	31
p	3520	3375	3220	3055	2880	2695	2500	2295

x	32	33	34	35	36	37	38	39
p	2080	1855	1620	1375	1120	855	580	295

and, finally, $p(40) = 0$. Answer: Drill ten new wells.

52. ROUND(x)=FLOOR$(x + \frac{1}{2})$. The range of ROUND(kx) is the set of all integers if k is nonzero; it is $\{0\}$ if $k = 0$.

55. ROUND4$(x) = \frac{1}{10000}$ROUND$(10000x)$.

58. 2.62

61. 0.72

64. 7.79

Section 1.2

1. $y = \frac{3}{2}x$ or $3x = 2y$.

4. The points $(2, 0)$ and $(0, -3)$ lie on L, so the slope of L is $\frac{3}{2}$. An equation of L is $2y = 3x - 6$.

7. The slope of L is -1; an equation is $y - 2 = 4 - x$.

10. The *other* line has slope $-\frac{1}{2}$, so L has slope 2 and therefore equation $y - 4 = 2(x + 2)$.

13. $(x + 1)^2 + (y + 1)^2 = 2^2$.

16. $x^2 + y^2 - \frac{2}{3}x - \frac{4}{3}y = \frac{11}{9}$: $x^2 - \frac{2}{3}x + \frac{1}{9} + y^2 - \frac{4}{3}y + \frac{4}{9} = \frac{16}{9}$; $(x - \frac{1}{3})^2 + (y - \frac{2}{3})^2 = (\frac{4}{3})^2$. Center $(\frac{1}{3}, \frac{2}{3})$, radius $\frac{4}{3}$.

19. $y - 3 = (x + 1)^2$: Opens upward, vertex at $(-1, 3)$.

22. $y = -(x^2 - x) = -(x^2 - x + \frac{1}{4}) + \frac{1}{4}$: $y - \frac{1}{4} = -(x - \frac{1}{2})^2$. Opens downward, vertex at $(\frac{1}{2}, \frac{1}{4})$.

25. $(x + 1)^2 + (y + 3)^2 = -10$: There are no points on the graph.

28. The graph is shown below.

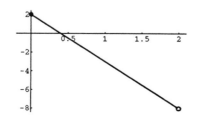

31. The graph is shown below.

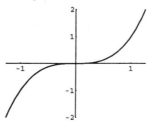

34. The graph is shown below.

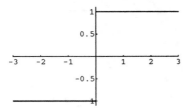

37. The graph is shown below.

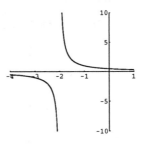

40. The graph is shown below.

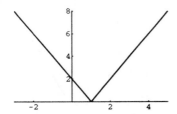

43. The graph is shown below.

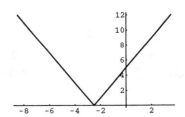

46. The graph is shown below.

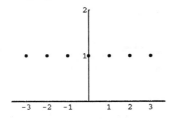

49. The graph is shown below.

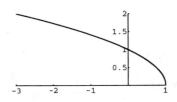

52. The graph is shown below.

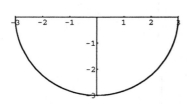

55. The graph is shown below.

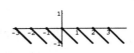

58. $(2.5, -1.5)$

61. $(2.25, 8.5)$

64. $\left(\frac{17}{9}, \frac{37}{9}\right)$

67. This is the same as Problem 66.

70. The function is given by:
$$f(x) = \begin{cases} 2x + 6 & \text{if } -3 \le x < -2; \\ 2 & \text{if } -2 \le x < 2; \\ \frac{1}{3}(10 - 2x) & \text{if } 2 \le x \le 5. \end{cases}$$

73. The graph is shown on the right.

$$x(t) = \begin{cases} 45t & \text{if } t < 1; \\ 45 + 75(t-1) & \text{if } 1 \le t. \end{cases}$$

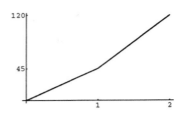

76. The graph is shown on the right.

$$x(t) = \begin{cases} 60t & \text{if } t < 0.5; \\ 30 - 60(t-0.5) & \text{if } 0.5 \le t < 1; \\ 60(t-1) & \text{if } 1 \le t. \end{cases}$$

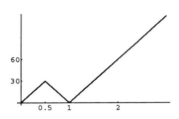

79. The graph is shown on the right.

$$C(x) = \begin{cases} 8 & \text{if } 0 < x \le 8; \\ 8 - 0.8[\![-(x-8)]\!] & \text{if } 8 < x \le 16. \end{cases}$$

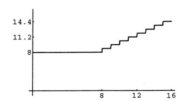

82. $T(t) = 61.25 + 17.85\cos(2\pi t/365)$. Average temperatures: $T(91) = 61.0$ on October 15 and $T(273) = 61.3$ on April 15. The graph is shown on the right.

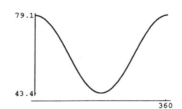

Section 1.3

1. $(f+g)(x) = x^2 + 3x - 2$, $(f \cdot g)(x) = x^3 + 3x^2 - x - 3$, and $(f/g)(x) = \dfrac{x+1}{x^2 + 2x - 3}$. The first two have domain the set $\mathbb{R}$ of all real numbers. The third has domain consisting of the set of all real numbers other than 1 and -3.

4. $(f+g)(x) = \sqrt{x+1} + \sqrt{5-x}$, $-1 \le x \le 5$; $(f \cdot g)(x) = \sqrt{5 + 4x - x^2}$, same domain.

$(f/g)(x) = \sqrt{\dfrac{x+1}{5-x}}$, $-1 \le x < 5$.

7. The graph is shown below.

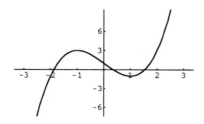

8. The graph is shown below.

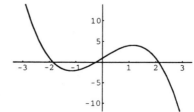

10. The graph is shown below.

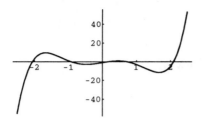

13. The graph is shown below.

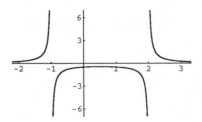

16. The graph is shown below.

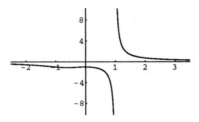

19. The graph is shown below.

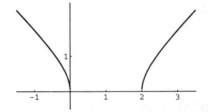

22. The graph is shown below.

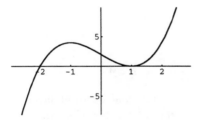

25. The graph is shown below.

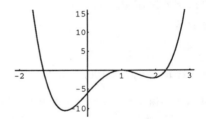

28. The graph is shown below.

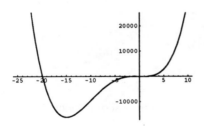

31. Graphs are shown below.

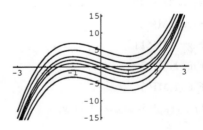

34. Graphs are shown below.

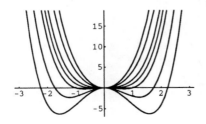

37. Graphs are shown below.

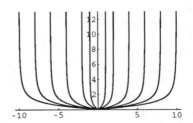

Section 1.4

1. Fig. 1.4.29   **4.** Fig. 1.4.32   **7.** Fig. 1.4.31   **10.** Fig. 1.4.30

16. $f(g(x)) = \sqrt{\cos x}$; $g(f(x)) = \cos \sqrt{x}$.

22. $k = 3$, $g(x) = 4 - x$

28. $k = -1$, $g(x) = 1 + x^2$ would seem to be the most natural answer.

31. one

34. three

37. three

40. six

43. 27.005 years

46. -0.7666

47. 4.84891

Chapter 1 Miscellaneous

1. $x \geq 4$

4. The set $\mathbb{R}$ of all real numbers

7. $x \leq \frac{2}{3}$

10. $[2, 4]$

13. First, $R = E/I = 100/I$, so $25 < 100/I < 50$; $\frac{1}{50} < I/100 < \frac{1}{25}$; $2 < I < 4$.

16. If the cylinder has radius and height x, then its volume is $V = \pi x^3$ and its total surface area is $S = 2\pi x^2 + 2\pi x^2 = 4\pi x^2$; but $x = (V/\pi)^{1/3}$, so $S = S(V) = 4\pi(V/\pi)^{2/3}$, $0 < V < \infty$.

19. $y - 5 = 2(x + 3)$

22. $2y = 3x - 12$

25. Fig. 1.MP.6

28. Fig. 1.MP.11

31. Fig. 1.MP.7

34. Fig. 1.MP.5

37. The graph is shown below.

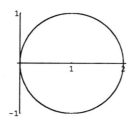

40. The graph is shown below.

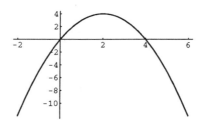

43. The graph is shown below.

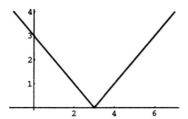

46. $|a| = |(a-b)+b| \le |a-b|+|b|$. Therefore $|a|-|b| \le |a-b|$.

49. $(x-4)(x+2) > 0$: Either $x > 4$ and $x > -2$ (so that $x > 4$) or $x < 4$ and $x < -2$ (so that $x < -2$). Answer: $(-\infty, -2) \cup (4, \infty)$.

52. -0.872 $\quad$ 4.205

55. -5.021, $\quad$ 0.896

58. $(5/3, 8/3)$

61. $(-33/16, 31/32)$

64. After shrinking, the tablecloth has dimensions $60 - x$ by $35 - x$. The area of this rectangle is 93% of the area of the original tablecloth, so $(60-x)(35-x) = (0.93)(35)(60)$. The larger solution of this quadratic equation is approximately 93.43, which we reject as too large. Answer: $x \approx 1.573$.

67. Three solutions.

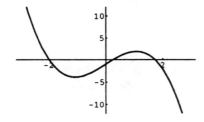

70. One solution.

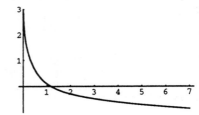

Chapter 2: Prelude to Calculus

Section 2.1

1. $f(x) = 0 \cdot x^2 + 0 \cdot x + 5$, so $m(a) = 0 \cdot 2 \cdot a + 0 \equiv 0$; $y \equiv 5$.

4. $m(a) = -4a$; $y = -8x + 9$.

7. Take $a = 2$, $b = -3$, and $c = 4$. Then $m(a) = 4a - 3$.

10. $f(x) = 15x - 3x^2$, so $m(a) = 15 - 6a$; $y = 3x + 12$.

13. $f(x) = 4x^2 + 1$, so $m(a) = 8a$; $y = 16x - 15$.

16. $m(a) = 10 - 2a$; horizontal tangent at $(5, 25)$.

19. $y = x - \dfrac{x^2}{100} = x - \frac{1}{100}x^2$, so $m(a) = 1 - \frac{1}{50}a$; thus there is a horizontal tangent at $(50, 25)$.

22. $y = x^2 - 10x + 25$, so $m(a) = 2a - 10$. Thus $m(x) = 0$ when $x = 5$, and when $x = 5$, $y = 0$. So there is a horizontal tangent at $(5, 0)$.

25. $m(x) = 2x$; the slope at P is -4; $y = -4x - 4$; $4y = x + 18$.

28. The tangent line in question has slope $2x_0$, and thus has equation $y - y_0 = 2x_0(x - x_0)$. Now $y_0 = (x_0)^2$, so this equation may be simplified to $y = 2x_0x - (x_0)^2$. The line with this equation meets the x-axis when $y = 0$, so that $2x_0x = (x_0)^2$. If $x_0 \neq 0$, this implies that $x = x_0/2$, and so—whether or not $x_0 = 0$—the tangent line meets the x-axis at $(x_0/2, 0)$.

31. If the two positive numbers are x and $50 - x$, then their product is $P(x) = x(50 - x)$, $0 < x < 50$. As in problem 30, the maximum possible value of P is $P(25) = 625$.

34. Let (a, b) denote the point where one of the two lines is tangent to the parabola. Then $b = 4a - a^2$, and the slope of the tangent line there is $4 - 2a$. As in the solution of Problem 33, we then have

$$4 - 2a = \frac{4a - a^2 - 5}{a - 2}.$$

The two solutions of this equation are $a = 1$ and $a = 3$. Thus one tangent line has slope 2 and the other has slope -2; their equations are $y - 5 = 2(x - 2)$ and $y - 5 = -2(x - 2)$ respectively.

37. 12

40. 0.25

Section 2.2

1. $\lim\limits_{x \to 3} (3x^2 + 7x - 12) = 3 \left(\lim\limits_{x \to 3} x\right)^2 + 7 \left(\lim\limits_{x \to 3} x\right) - \lim\limits_{x \to 3} 12 = 3 \cdot 3^2 + 7 \cdot 3 - 12 = 36.$

4. $\lim\limits_{x \to -2} (x^3 - 3x + 3)(x^2 + 2x + 5) = \lim\limits_{x \to -2} (x^3 - 3x + 3) \cdot \lim\limits_{x \to -2} (x^2 + 2x + 5) = 1 \cdot 5 = 5.$

7. $\lim\limits_{x \to 3} \dfrac{(x^2 + 1)^3}{(x^3 - 25)^3} = \dfrac{\lim\limits_{x \to 3} (x^2 + 1)^3}{\lim\limits_{x \to 3} (x^3 - 25)^3} = \dfrac{\left(\lim\limits_{x \to 3} (x^2 + 1)\right)^3}{\left(\lim\limits_{x \to 3} (x^3 - 25)\right)^3} = \dfrac{10^3}{2^3} = \dfrac{1000}{8} = 125.$

10. $\lim\limits_{y \to 4} \sqrt{27 - \sqrt{y}} = \sqrt{\lim\limits_{y \to 4} (27 - \sqrt{y})} = \sqrt{\sqrt{25}} = 5.$

13. $\lim\limits_{z \to 8} \dfrac{z^{2/3}}{z - \sqrt{2z}} = \dfrac{\lim\limits_{z \to 8} z^{2/3}}{\lim_{z \to 8} (z - \sqrt{2z})} = \frac{4}{4} = 1.$

16. $\lim\limits_{t \to -4} \sqrt[3]{(t + 1)^6} = \sqrt[3]{\lim\limits_{t \to -4} (t + 1)^6} = 9.$

19. $\displaystyle\lim_{x\to-1}\frac{x+1}{x^2-x-2}=\lim_{x\to-1}\frac{x+1}{(x+1)(x-2)}=\lim_{x\to-1}\frac{1}{x-2}=-\frac{1}{3}.$

22. $\displaystyle\lim_{y\to-1/2}\frac{4y^2-1}{4y^2+8y+3}=\lim_{y\to-1/2}\frac{(2y-1)(2y+1)}{(2y+3)(2y+1)}=\lim_{y\to-1/2}\frac{2y-1}{2y+3}=-\frac{2}{2}=-1.$

25. $\displaystyle\lim_{z\to-2}\frac{(z+2)^2}{z^4-16}=\lim_{z\to-2}\frac{(z+2)(z+2)}{(z+2)(z-2)(z^2+4)}=\lim_{z\to-2}\frac{z+2}{(z-2)(z^2+4)}=0.$

28. $\displaystyle\lim_{y\to-3}\frac{y^3+27}{y^2-9}=\lim_{y\to-3}\frac{(y+3)(y^2-3y+9)}{(y+3)(y-3)}=\lim_{y\to-3}\frac{y^2-3y+9}{y-3}=-\frac{27}{6}=-\frac{9}{2}.$

31. $\displaystyle\lim_{x\to4}\frac{x-4}{\sqrt{x}-2}=\lim_{x\to4}\frac{(\sqrt{x}-2)(\sqrt{x}+2)}{\sqrt{x}-2}=\lim_{x\to4}\sqrt{x}+2=4.$

34. $\displaystyle\lim_{h\to0}\frac{1}{h}\left(\frac{1}{\sqrt{9+h}}-\frac{1}{3}\right)=\lim_{h\to0}\frac{3-\sqrt{9+h}}{3h\sqrt{9+h}}$

$$=\lim_{h\to0}\left(\frac{3-\sqrt{9+h}}{3h\sqrt{9+h}}\right)\cdot\left(\frac{3+\sqrt{9+h}}{3+\sqrt{9+h}}\right)$$

$$=\frac{9-(9+h)}{3h\sqrt{9+h}\,(3+\sqrt{9+h}\,)}=\lim_{h\to0}\frac{-1}{3\sqrt{9+h}\,(3+\sqrt{9+h}\,)}=-\frac{1}{54}.$$

37. $\displaystyle\frac{f(x+h)-f(x)}{h}=\frac{(x+h)^3-x^3}{h}=\frac{x^3 x^2 h+3xh^2+h^3-x^3}{h}=3x^2+3xh+h^2\to3x^2$ as $h\to0$.

When $x=2$, $y=f(2)=x^3=8$ and the slope of the tangent line to this curve at $x=2$ is $3x^2=12$, so an equation of this tangent line is $y=12x-16$.

40. $\displaystyle\frac{f(x+h)-f(x)}{h}=\frac{\left(\dfrac{1}{x+h+1}\right)-\left(\dfrac{1}{x+1}\right)}{h}=\frac{x+1-x-h-1}{h(x+1)(x+h+1)}=\frac{-1}{(x+1)(x+h+1)}.$

This approaches $-\dfrac{1}{(x+1)^2}$ as h approaches 0. When $x=2$, $y=f(2)=\frac{1}{3}$ and the slope of the line tangent to this curve at $x=2$ is $-\frac{1}{9}$, so an equation of this tangent line is $y-\frac{1}{3}=-\frac{1}{9}(x-2)$, or $y=-\frac{1}{9}(x-5)$.

43. $\displaystyle\frac{f(x+h)-f(x)}{h}=\frac{\left(\dfrac{1}{\sqrt{x+h+2}}\right)-\left(\dfrac{1}{\sqrt{x+2}}\right)}{h}$

$$=\left(\frac{\sqrt{x+2}-\sqrt{x+h+2}}{h\sqrt{x+2}\sqrt{x+h+2}}\right)\cdot\left(\frac{\sqrt{x+2}+\sqrt{x+h+2}}{\sqrt{x+2}+\sqrt{x+h+2}}\right)$$

$$=\frac{-h}{h\sqrt{x+2}\sqrt{x+h+2}\,(\sqrt{x+2}+\sqrt{x+h+2}\,)}\to\frac{-1}{(x+2)\,(2\sqrt{x+2}\,)}\text{ as }h\to0.$$

When $x=2$, $y=f(2)=\frac{1}{2}$ and the slope of the line tangent to this curve at $x=2$ is $\frac{-1}{16}$, so an equation of this tangent line is $y-\frac{1}{2}=-\frac{1}{16}(x-2)$, or $y=\frac{-1}{16}(x-10)$.

46. $\displaystyle\frac{f(x+h)-f(x)}{h}=\frac{\left(\dfrac{(x+h)^2}{x+h+1}\right)-\left(\dfrac{x^2}{x+1}\right)}{h}=\frac{(x+1)(x+h)^2-(x+h+1)(x^2)}{h(x+1)(x+h+1)}$

$$=\frac{x^2+xh+2x+h}{(x+1)(x+h+1)}\to\frac{x^2+2x}{(x+1)^2}\text{ as }h\to0.$$

When $x=2$, $y=f(2)=\frac{4}{3}$ and the slope of the line tangent to this curve at $x=2$ is $\frac{8}{9}$, so an equation of this tangent line is $y-\frac{4}{3}=\frac{8}{9}(x-2)$, or $9y=8x-4$.

49.

x	10^{-2}	10^{-4}	10^{-6}	10^{-8}	10^{-10}
$f(x)$	0.16662	0.166666	0.166667	0.166667	0.166667

x	-10^{-2}	-10^{-4}	-10^{-6}	-10^{-8}	-10^{-10}
$f(x)$	0.166713	0.166667	0.166667	0.166667	0.166667

The limit appears to be $\frac{1}{6}$.

52.

x	10^{-2}	10^{-4}	10^{-6}	10^{-8}	10^{-10}
$f(x)$	-0.222225	-0.222222	-0.222222	-0.222225	-0.222222

x	-10^{-2}	-10^{-4}	-10^{-6}	-10^{-8}	-10^{-10}
$f(x)$	-0.222225	-0.222222	-0.222222	-0.222222	-0.222222

Limit: $-\frac{2}{9}$.

55.

x	10^{-2}	10^{-3}	10^{-4}	10^{-5}	10^{-6}
$f(x)$	0.166666	0.166667	0.166667	0.166667	0.166667

x	-10^{-2}	-10^{-3}	-10^{-4}	-10^{-5}	-10^{-6}
$f(x)$	0.166666	0.166667	0.166667	0.166667	0.166667

The limit appears to be $\frac{1}{6}$.

61. $\sin\left(\dfrac{\pi}{2^{-n}}\right) = \sin\left(2\pi \cdot 2^{(n-1)}\right) = 0$ for every positive integer n. Therefore $\lim\limits_{x\to 0}\sin\left(\dfrac{\pi}{x}\right)$, if it were to exist, would be 0. Notice however that $\sin\left(3^n \cdot \dfrac{\pi}{2}\right)$ alternates between $+1$ and -1 for $n = 1, 2, 3, \dots$. Therefore $\lim\limits_{x\to 0}\sin\left(\dfrac{\pi}{x}\right)$ does not exist.

Section 2.3

1. $\theta \cdot \dfrac{\theta}{\sin\theta} \to 0 \cdot 1 = 0$ as $\theta \to 0$.

4. $\dfrac{\tan\theta}{\theta} = \dfrac{\sin\theta}{\theta\cos\theta} \to 1$ as $\theta \to 0$.

7. Let $z = 5x$. Then $z \to 0$ as $x \to 0$, and $\dfrac{\sin 5x}{x} = \dfrac{5\sin z}{z} \to 5$.

10. Replace $1 - \cos 2x$ with $1 - (\cos^2 x - \sin^2 x) = 2\sin^2 x$ to see that the limit is zero.

13. Multiply numerator and denominator by $1 + \cos x$.

16. $\lim\limits_{\theta\to 0}\dfrac{\sin 2\theta}{\theta} = \lim\limits_{\theta\to 0}\dfrac{2\sin\theta\cos\theta}{\theta} = 2\cdot 1 \cdot 1 = 2$.

19. $\lim\limits_{z\to 0}\dfrac{\tan z}{\sin 2z} = \lim\limits_{z\to 0}\dfrac{\sin z}{2\sin z \cos^2 z} = \frac{1}{2}$.

22. $\lim\limits_{x\to 0}\dfrac{x - \tan x}{\sin x} = \lim\limits_{x\to 0}\dfrac{\left(\dfrac{x}{x} - \dfrac{\tan x}{x}\right)}{\left(\dfrac{\sin x}{x}\right)} = \dfrac{1 - 1}{1} = 0$.

28. Because $-1 \le \sin\dfrac{1}{x} \le 1$ for all $x \ne 0$, $-\left|\sqrt[3]{x}\right| \le \sqrt[3]{x}\sin\dfrac{1}{x} \le \left|\sqrt[3]{x}\right|$ for all $x \ne 0$. Now let $x \to 0$ to obtain the limit zero.

31. If $x < 1$, then $x - 1 < 0$, so the limit does not exist.

34. If $x > 3$ then $9 - x^2 < 0$, so the limit does not exist.

37. $\dfrac{4x}{x-4} \to +\infty$ as $x \to 4^+$, so the limit does not exist. In such a case it is also correct to write

$$\lim_{x \to 4^+} \sqrt{\dfrac{4x}{x-4}} = +\infty.$$

40. If $0 > x > -4$, then $16 - x^2 > 0$, so $\dfrac{16 - x^2}{\sqrt{16 - x^2}} = \sqrt{16 - x^2} \to 0$ as $x \to -4^+$.

43. If $x > 2$ then $x - 2 > 0$, so $\dfrac{2 - x}{|x - 2|} = -1$. Therefore the limit is also -1.

46. If $x < 0$ then $x - |x| = 2x$, so the limit is $1/2$.

49. $f(x) \to +\infty$ as $x \to 1^+$, $f(x) \to -\infty$ as $x \to 1^-$.

52. If x is near 5 then $2x - 5$ is near 5. So $f(x) \to +\infty$ as $x \to 5^-$, $f(x) \to -\infty$ as $x \to 5^+$.

55. If $x > 1$ then $f(x) = \dfrac{1}{x-1}$; if $x < 1$ then $f(x) = \dfrac{1}{1-x}$. Therefore $f(x) \to +\infty$ as $x \to 1$.

58. $\dfrac{x-1}{x^2 - 3x + 2} = \dfrac{x-1}{(x-1)(x-2)} = \dfrac{1}{x-2}$ for $x \neq 1$, $x \neq 2$. So $f(x) \to -1$ as $x \to 1$, $f(x) \to +\infty$ as $x \to 2^+$, and $f(x) \to -\infty$ as $x \to 2^-$.

61. If x is an even integer then $f(x) = 3$, if x is an odd integer then $f(x) = 1$, and $\lim_{x \to a} f(x) = 2$ for all real number values of a.

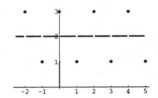

64. If n is any odd integer $\lim_{x \to n^-} f(x) = 1$ and $\lim_{x \to n^+} f(x) = -1$; if n is any even integer $\lim_{x \to n^-} f(x) = -1$ and $\lim_{x \to n^+} f(x) = 1$. Note: $\lim_{x \to a} f(x)$ exists if and only if a is not an integer. The graph is shown on the right.

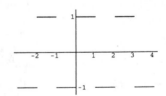

67. If n is any integer, $\lim_{x \to n^-} f(x) = -1$ and $\lim_{x \to n^+} f(x) = -1$. Note: $\lim_{x \to a} f(x) = -1$ at every real value of a. The graph is shown on the right.

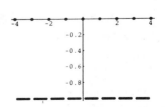

70. Let $f(x) = \text{sgn}(x)$ and $g(x) = -\text{sgn}(x)$.

73. $\lim\limits_{x \to 0} f(x) = 1$. The graph is shown on the right.

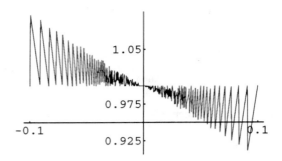

Section 2.4

4.
$$\lim_{x \to a} g(x) = \frac{\lim_{x \to a} x^3}{\lim_{x \to a} x^2 + 2\lim_{x \to a} x + 5} = \frac{(\lim_{x \to a} x)^3}{(\lim_{x \to a} x)^2 + 2\lim_{x \to a} x + 5} = \frac{a^3}{a^2 + 2a + 5} = g(a).$$
Therefore $\lim\limits_{x \to a} g(x)$ exists and is equal to $g(a)$ at every real value of a, and so g is continuous at every real number.

16. The domain is the set of all real numbers other than zero. The function g is continuous on its domain (it is not continuous at $x = 0$ because it is not defined there).

19. f is continuous on its domain $\mathbb{R}$ (the denominator is never zero).

22. h is continuous on $\mathbb{R}$ (the denominator is never zero).

25. Every real number has a [unique] cube root, so by Theorem 2 and the limit laws f is continuous on its domain, $x \neq 1$.

28. f is continuous on its domain, the interval $[-3, 3]$.

31. f is continuous on its domain, the set of all real numbers other than zero.

34. The domain of f consists of those values of x for which $\sin x \geq 0$. So its domain is the union of all intervals of the form $[n\pi, (n+1)\pi]$ where n is an even integer. It is continuous on its domain.

37. The function f is discontinuous at $x = -3$, and cannot be made continuous there because it has no limit there.

40. $G(u) = \dfrac{u+1}{u^2 - u - 6} = \dfrac{u+1}{(u-3)(u+2)}$, so G has no limit at either $u = 3$ or $u = -2$. Therefore it is not continuous at either point nor can it be made continuous at either point.

43. Because $f(x) \to 1$ as $x \to 17^+$ whereas $f(x) \to -1$ as $x \to 17^-$, f has no limit at $x = 17$ and cannot be made continuous there.

46. The function f is not continuous at $x = 1$ because it is not defined there. If we define $f(1)$ to be 2, then f will become continuous at $x = 1$ as well because its limit at $x = 1$ will be equal to its value there.

49. If f is continuous at $x = 0$ then $f(0) = c = \lim_{x \to 0+} f(x) = 4 - 0^2$, and so $c = 4$.

52. If f is continuous at $x = \pi$ then $f(\pi) = c^3 - \pi^3 = \lim_{x \to 3+} f(x) = c \sin \pi = 0$, and so $c = \pi$.

55. Let $f(x) = x^3 - 3x^2 + 1$. Then f is continuous on $[0, 1]$ because $f(x)$ is a polynomial. Moreover, $f(0) = 1 > 0 > -1 = f(1)$. Therefore $f(x) = 0$ for some number x in $[0, 1]$ ($x \approx 0.6527$).

58. Let $f(x) = x^5 - 5x^3 + 3$. Then f is continuous on $[-3, -2]$, because $f(x)$ is a polynomial. Moreover, $f(-3) = -105 < 0$ and $f(-2) = 11 > 0$. Therefore $f(x) = 0$ for some number x in $[-3, -2]$. (The value of that number x is approximately -2.291164).

61. S is discontinuous at the end of each year; that is, at each integral value of t.

The graph is shown on the right.

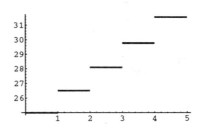

Chapter 2 Miscellaneous

1. $0^2 - 3 \cdot 0 + 4 = 4$

4. $(1 + 1 - 1)^{17} = 1$

7. $\dfrac{x^2 - 1}{1 - x} = -(x + 1) \to -2$ as $x \to 1$.

10. $\dfrac{4 - x^2}{3 + x} \to \frac{4}{3}$ as $x \to 0$.

13. $16^{3/4} = 8$

16. $x - \sqrt{x^2 - 1} = \dfrac{x^2 - (x^2 - 1)}{x + \sqrt{x^2 - 1}} = \dfrac{1}{x + \sqrt{x^2 - 1}} \to 1$ as $x \to 1^+$.

19. $\sqrt{(2 - x)^2} = |2 - x| = x - 2$ for $x > 2$, so the fraction and its limit are equal to -1.

22. If $x < 3$ then $x^2 - 9 < 0$, so the limit does not exist.

25. The numerator approaches 4 while the denominator approaches zero through positive values. Therefore the limit is $+\infty$.

28. If $x \neq 1$ then the fraction is equal to $\dfrac{1}{x - 1}$, and its denominator approaches zero through negative values, so the limit is $-\infty$.

31. $\dfrac{\sin 3x}{x} = \dfrac{3 \sin 3x}{3x} \to 3$ as $x \to 0$.

34. $\lim\limits_{x \to 0} \dfrac{\tan 2x}{\tan 3x} = \lim\limits_{x \to 0} \dfrac{2}{3} \cdot \dfrac{\sin 2x}{2x} \cdot \dfrac{3x}{\sin 3x} \cdot \dfrac{\cos 3x}{\cos 2x} = \frac{2}{3}$.

37. $\lim\limits_{x \to 0} \dfrac{1 - \cos 3x}{2x^2} = \lim\limits_{x \to 0} \dfrac{1 - \cos^2 3x}{2x^2(1 + \cos 3x)} = \lim\limits_{x \to 0} \dfrac{9}{2(1 + \cos 3x)} \cdot \dfrac{\sin 3x}{3x} \cdot \dfrac{\sin 3x}{3x} = \frac{9}{4}$.

40. $\lim\limits_{x \to 0} x^2 \cot^2 3x = \lim\limits_{x \to 0} \dfrac{3x}{\sin 3x} \cdot \dfrac{3x}{\sin 3x} \cdot \dfrac{\cos^2 3x}{9} = \frac{1}{9}$.

43. $f'(x) = 6x + 4$; $f'(1) = 10$: $f(1) = 2$; $y - 2 = 10(x - 1)$.

46. $f(x) = \frac{1}{3}x - \frac{1}{16}x^2$, so $f'(x) = \frac{1}{3} - \frac{1}{8}x$. Then $f(1) = \frac{13}{48}$ and $f'(1) = \frac{5}{24}$, so one equation of the line is $y - \frac{13}{48} = \frac{5}{24}(x - 1)$.

49. $f'(x) = \lim\limits_{h \to 0} \dfrac{\dfrac{1}{3 - x - h} - \dfrac{1}{3 - x}}{h}$

$= \lim\limits_{h \to 0} \dfrac{3 - x - 3 + x + h}{h(3 - x - h)(3 - x)}$

$= \lim\limits_{h \to 0} \dfrac{1}{(3 - x - h)(3 - x)} = \dfrac{1}{(3 - x)^2}.$

52. $f'(x) = \lim\limits_{h \to 0} \dfrac{\dfrac{x+h}{x+h+1} - \dfrac{x}{x+1}}{h}$

$= \lim\limits_{h \to 0} \dfrac{x^2 + xh + x + h - x^2 - xh - x}{h(x+h+1)(x+1)} = \lim\limits_{h \to 0} \dfrac{1}{(x+h+1)(x+1)} = \dfrac{1}{(x+1)^2}.$

55. The slope of a line tangent to $y = x^2$ at (a, a^2) is $2a$. If such a line passes through $(3, 4)$, then the two-point slope formula yields $2a = (a^2 - 4)/(a - 3)$. It is now easy to solve to obtain the two values $a = 3 \pm \sqrt{5}$.

58. Every rational function is continuous wherever its denominator is nonzero. Here the denominator of $f(x)$ is zero when $x = 2$, and f cannot be made continuous there.

61. Let $f(x) = x^5 + x - 1$. Then $f(0) = -1 < 0 < 1 = f(1)$. Because $f(x)$ is a polynomial, it is continuous on $[0, 1]$, so f has the intermediate value property there. Hence there exists a number c in $(0, 1)$ such that $f(c) = 0$. Thus $c^5 + c - 1 = 0$, and so the equation $x^5 + x - 1 = 0$ has a solution. (The value of c is approximately 0.754877666.)

64. Let $h(x) = x + \tan x$. Then $h(\pi) = \pi > 0$ and $\lim\limits_{x \to (\pi/2)+} h(x) = -\infty$. The latter fact implies that $h(r) < 0$ for some number r slightly larger than $\pi/2$. Because h is continuous on the interval $[r, \pi]$, h has the intermediate value property there, so $h(c) = 0$ for some number c between r and π, and thus between $\pi/2$ and π.

Chapter 3: The Derivative

Section 3.1

1. $f'(x) = 4$ $\quad$ $(a = 0, b = 4, c = -5)$

4. $f'(x) = -49$

10. $\dfrac{du}{dt} = 14t + 13$

13. $f'(x) = \lim\limits_{h \to 0} \dfrac{(x+h)^2 + 5 - x^2 - 5}{h} = \lim\limits_{h \to 0} \dfrac{2xh + h^2}{h} = \lim\limits_{h \to 0} (2x + h) = 2x.$

16. $f'(x) = \lim\limits_{h \to 0} \dfrac{\dfrac{1}{3 - x - h} - \dfrac{1}{3 - x}}{h} = \lim\limits_{h \to 0} \dfrac{3 - x - 3 + x + h}{h(3 - x - h)(3 - x)}$

$\qquad = \lim\limits_{h \to 0} \dfrac{1}{(3 - x - h)(3 - x)} = \dfrac{1}{(3 - x)^2}.$

19. $f'(x) = \lim\limits_{h \to 0} \dfrac{\dfrac{x + h}{1 - 2x - 2h} - \dfrac{x}{1 - 2x}}{h} = \lim\limits_{h \to 0} \dfrac{(x + h)(1 - 2x) - x(1 - 2x - 2h)}{h(1 - 2x - 2h)(1 - 2x)}$

$\qquad = \lim\limits_{h \to 0} \dfrac{1}{(1 - 2x - 2h)(1 - 2x)} = \dfrac{1}{(1 - 2x)^2}.$

22. $\dfrac{dx}{dt} = -32t + 160$; $\quad \dfrac{dx}{dt} = 0$ when $t = 5$; $\quad x(5) = 425$

25. $\dfrac{dx}{dt} = -20 - 10t$; $\quad \dfrac{dx}{dt} = 0$ when $t = -2$; $\quad x(-2) = 120$

28. $\dfrac{dy}{dt} = -32t + 128$; $\quad \dfrac{dy}{dt} = 0$ when $t = 4$; $\quad y(4) = 281$ (ft)

31. Matches (e).

34. Matches (a).

37. Let r denote the radius of the circle. Then $A = \pi r^2$ and $C = 2\pi r$, so $r = \dfrac{C}{2\pi}$. Thus $A = \dfrac{1}{4\pi} C^2$, so the rate of change of A with respect to C is $dA/dC = \dfrac{C}{2\pi}$.

40. Because $V(t) = 10 - \frac{1}{5}t + \frac{1}{1000}t^2$, $V'(t) = -\frac{1}{5} + \frac{1}{500}t$ and the rate at which the water is leaking out one minute later ($t = 60$) is $V(60) = -\frac{2}{25}$ (gal/s) or—if you prefer— -4.8 gal/min. The average rate of change of V from $t = 0$ until $t = 100$ is $\dfrac{V(100) - V(0)}{100 - 0} = \dfrac{0 - 10}{100} = -\dfrac{1}{10}$. The instantaneous rate of change of V will have this value when $V'(t) = -\frac{1}{10}$, which we easily solve for $t = 50$.

43. On our graph, the tangent line at the point $(20, 810)$ has slope $m_1 \approx 0.6$ and the tangent line at $(40, 2686)$ has slope $m_2 \approx 0.9$. A line of slope 1 on our graph corresponds to a velocity of 125 ft/s (because the line through $(0, 0)$ and $(10, 1250)$ has slope 1), and thus we estimate the velocity of the car at time $t = 20$ to be about $(0.6)(125) = 75$ ft/s, and at time $t = 40$ it is traveling at about $(0.9)(125) = 112.5$ ft/s. The method is crude; the answer in the back of the textbook is quite different simply because it was obtained by someone else.

46. A right circular cylinder of radius r and height h has volume $V = \pi r^2 h$ and total surface area S obtained by adding the areas of its top, bottom, and curved side: $S = 2\pi r^2 + 2\pi rh$. We are given $h = 2r$, so $V(r) = 2\pi r^3$ and $S(r) = 6\pi r^2$. Also $dV/dr = 6\pi r^2 = S(r)$, so the rate of change of volume with respect to radius is indeed equal to total surface area.

49. Let $V(t)$ denote the volume (in cm^3) of the snowball at time t (in hours), and let $r(t)$ denote its

radius then. From the data given in the problem, $r = 12 - t$. The volume of the snowball is

$$V = \frac{4}{3}\pi r^3 = \frac{4}{3}\pi(12 - t)^3 = \frac{4}{3}\pi\left(1728 - 432t + 36t^2 - t^3\right),$$

so its instantaneous rate of change is

$$V'(t) = \frac{4}{3}\pi\left(-432 + 72t - 3t^2\right).$$

Hence its rate of change of volume when $t = 6$ is $V'(6) = -144\pi$ cm^3/h. Its average rate of change of volume from $t = 3$ to $t = 9$ in cm^3/h is

$$\frac{V(9) - V(3)}{9 - 3} = \frac{36\pi - 972\pi}{6} = -156\pi \ (\text{cm}^3/\text{h}).$$

52. Because $P(t) = 100 + 4t + \frac{3}{10}t^2$, we have $P'(t) = 4 + \frac{3}{5}t$. The year 1986 corresponds to $t = 6$, so the rate of change of P then was $P'(6) = 7.6$ (thousands per year). The average rate of change of P from 1983 ($t = 3$) to 1988 ($t = 8$) was

$$\frac{P(8) - P(3)}{8 - 3} = \frac{151.2 - 114.7}{5} = 7.3 \ (\text{thousands per year}).$$

Section 3.2

1. $f'(x) = 6x - 1$

4. $g'(x) = (2x^2 - 1)(3x^2) + (4x)(x^3 + 2) = 10x^4 - 3x^2 + 8x$

7. $f(y) = 4y^3 - y$: $\quad f'(y) = 12y^2 - 1.$

10. $f'(t) = \dfrac{0 - (1)(-2t)}{(4 - t^2)^2} = \dfrac{2t}{(4 - t^2)^2}$

13. By first multiplying the two factors of $g(t)$:

$$g'(t) = 5t^4 + 4t^3 + 3t^2 + 4t.$$

By applying the product rule:

$$g'(t) = (t^2 + 1)(3t^2 + 2t) + (t^3 + t^2 + 1)(2t).$$

Of course both answers are the same.

16. $f(x) = 2x - 3 + 4x^{-1} - 5x^{-2}$, so $f'(x) = 2 - \dfrac{4}{x^2} + \dfrac{10}{x^3} = \dfrac{2x^3 - 4x + 10}{x^3}.$

22. $h'(x) = \dfrac{(2x - 5)(6x^2 + 2x - 3) - (2)(2x^3 + x^2 - 3x + 17)}{(2x - 5)^2} = \dfrac{8x^3 - 28x^2 - 10x - 19}{(2x - 5)^2}$

25. First rewrite the function in the form $g(x) = \dfrac{x^3 - 2x^2}{2x - 3}$.

Then $g'(x) = \dfrac{(2x - 3)(3x^2 - 4x) - (2)(x^3 - 2x^2)}{(2x - 3)^2} = \dfrac{4x^3 - 13x^2 + 12x}{(2x - 3)^2}.$

28. $\dfrac{dx}{dt} = -3t^{-2} + 8t^{-3}$

34. $u = \dfrac{t^2}{t^2 - 4}$ for $t \neq 0$. So $\dfrac{du}{dt} = \dfrac{(t^2 - 4)(2t) - (t^2)(2t)}{(t^2 - 4)^2} = -\dfrac{8t}{(t^2 - 4)^2}.$

37. $y = \dfrac{10x^6}{15x^5 - 4}$ for $x \neq 0.$

40. $h(w) = w^{-1} + 10w^{-2}$, so $h'(w) = -w^{-2} - 20w^{-3}$.

43. $\dfrac{dy}{dx} = -(x-1)^2$; the slope at P is -1. An equation of the tangent line is $y - 1 = -(x-2)$.

46. $y = \dfrac{x^2}{x-1}$ for $x \neq 0$, so

$$\frac{dy}{dx} = \frac{(x-1)(2x) - x^2}{(x-1)^2} = \frac{x^2 - 2x}{(x-1)^2}.$$

The slope at P is zero; the line has equation $y = 4$.

49. $\dfrac{dy}{dx} = \dfrac{3x^2 + 6x}{(x^2 + x + 1)^2}$

52. $W = \dfrac{2 \times 10^9}{R^2} = (2 \times 10^9)R^{-2}$, so $\dfrac{dW}{dR} = -\dfrac{4 \times 10^9}{R^3}$; when $R = 3960$, $\dfrac{dW}{dR} = -\dfrac{62500}{970299}$ lb/mi. Thus W decreases initially at about 1.03 oz/mi.

55. The slope of the tangent line can be computed using dy/dx at $x = a$ and also by using the two points known to lie on the line. We thereby find that $3a^2 = \dfrac{a^3 - 5}{a - 1}$. This leads to the equation $(a+1)(2a^2 - 5a + 5) = 0$. The quadratic factor has negative discriminant, so the only real solution of the cubic equation is $a = -1$. The point of tangency is $(-1, -1)$, the slope there is 3, and the equation of the line in question is $y = 3x + 2$.

58. Let $(a, 1/a)$ be a point of tangency. The slope of the tangent there is $-1/a^2$, so $-1/a^2 = -2$. Thus there are two possible values for a: $\pm\frac{1}{2}\sqrt{2}$. These lead to the equations of the two lines: $y = -2x + 2\sqrt{2}$ and $y = -2x - 2\sqrt{2}$.

61. $D_x[f(x)]^3 = f'(x)f(x)f(x) + f(x)f'(x)f(x) + f(x)f(x)f'(x) = 3[f(x)]^2 f'(x)$.

64. With $f(x) = x^2 + x + 1$ and $n = 100$, we obtain $D_x(x^2 + x + 1)^{100} = 100(x^2 + x + 1)^{99}(2x + 1)$.

67. When $n = 0$, $\dfrac{dy}{dx} = -\dfrac{2x}{(1 + x^2)^2}$; when $n = 2$, $\dfrac{dy}{dx} = \dfrac{2x}{(1 + x^2)^2}$. In both cases there is only one horizontal tangent, at the point where $x = 0$.

70. $f'(x) = 1$ when $x = \pm 1$.

Section 3.3

1. $dy/dx = 5(3x + 4)^4(3) = 15(3x + 4)^4$

4. $y = (2x + 1)^{-3}$: $dy/dx = -6(2x + 1)^{-4}$

7. $dy/dx = -4(2 - x)^3(3 + x)^7 + 7(2 - x)^4(3 + x)^6$

10. $dy/dx = \dfrac{-6x(4 + 5x + 6x^2)(1 - x^2)^2 - 2(5 + 12x)(1 - x^2)^3}{(4 + 5x + 6x^2)^3}$.

A common factor of $4 + 5x + 6x^2$ was cancelled from both terms in the numerator and from the denominator.

13. $dy/dx = 3(u + 1)^2 du/dx = -\dfrac{6}{x^3}\left(\dfrac{1}{x^2} + 1\right)^2$

16. $dy/dx = \dfrac{-15}{(3x - 2)^6}$

19. $dy/dx = -\dfrac{2(x^3 - 1)^2(x^3 + 1)^2(5x^6 - 14)}{x^{29}}$. This is the result of considerable simplification of the initial result, which was $dy/dx = -4x^{-5}(x^{-2} - x^{-8})^3 + 3x^{-4}(8x^{-9} - 2x^{-3})(x^{-2} - x^{-8})^2$.

22. $f'(x) = -\dfrac{15x^2}{(5x^3 + 2)^2}$

28. $h'(z) = 6z^3(z^2 + 4)^2 + 2z(z^2 + 4)^3$

31. $f'(u) = 8u(u + 1)^3(u^2 + 1)^3 + 3(u + 1)^2(u^2 + 1)^4$

34. $p'(t) = \dfrac{4(t^{-2} + 2t^{-3} + 3t^{-4})}{(t^{-1} + t^{-2} + t^{-3})^5} = \dfrac{4t^{11}(t^2 + 2t + 3)}{(t^2 + t + 1)^5}$

40. $dy/dx = -3(1 - x)^2$

43. $dy/dx = -\dfrac{2x}{(x^2 + 1)^2}$

46. $g'(t) = 3\sin^2 t \cos t$

49. $r(t) = 2t$ and $a(t) = \pi(2t)^2$ and $a'(t) = 8\pi t$. When $r = 10$, $t = 5$, and at that time the rate of change of area with respect to time is $a'(5) = 40\pi$.

52. Let x denote the length of each side of the triangle. Then its altitude is $\frac{1}{2}x\sqrt{3}$, and so its area is $A = \frac{1}{4}x^2\sqrt{3}$. Therefore the rate of change of its area with respect to time t (in seconds) is

$$\frac{dA}{dt} = \frac{1}{2}x\sqrt{3}\,\frac{dx}{dt}.$$

We are given $x = 10$ and $dx/dt = 2$, so at that point the area is increasing at $10\sqrt{3}$ in.^{2}/s.

55. $G'(t) = f'(h(t)) \cdot h'(t)$. Now $h(1) = 4$, $h'(1) = -6$, and $f'(4) = 3$, so $G'(1) = 3 \cdot (-6) = -18$.

58. Let V denote the volume of the balloon and r its radius at time t (in seconds). We are given $dV/dt = 200\pi$. Now

$$\frac{dV}{dt} = \frac{dV}{dr} \cdot \frac{dr}{dt} = 4\pi r^2 \frac{dr}{dt}.$$

When $r = 5$, we have $200\pi = 4\pi \cdot 25 \cdot (dr/dt)$, so $dr/dt = 2$. Answer: When $r = 5$ (cm), the radius of the balloon is increasing at 2 cm/s.

61. Let V denote the volume of the snowball and A its surface area at time t (in hours). Then

$$dV/dt = kA \text{ and } A = cV^{2/3}$$

(the latter because A is proportional to r^2, whereas V is proportional to r^3). Therefore

$$dV/dt = \alpha V^{2/3} \text{ and thus } dt/dV = \beta V^{-2/3}$$

(α and β are constants). From the last equation we may conclude that $t = \gamma V^{1/3} + \delta$ for some constants γ and δ, so that $V = V(t) = (Pt + Q)^3$ for some constants P and Q. From the information $500 = V(0) = Q^3$ and $250 = V(1) = (P + Q)^3$, we find that $Q = 5\sqrt[3]{4}$ and that $P = -5 \cdot (\sqrt[3]{4} - \sqrt[3]{2})$. Now $V(t) = 0$ when $PT + Q = 0$; it turns out that

$$T = \frac{\sqrt[3]{2}}{\sqrt[3]{2} - 1} \approx 4.8473.$$

Therefore the snowball finishes melting at about 2:50:50 P.M. on the same day.

Section 3.4

1. $f(x) = 4x^{5/2} + 2x^{-1/2}$, so $f'(x) = 10x^{3/2} - x^{-3/2}$.

4. $h(z) = (7 - 6z)^{-1/3}$, so $h'(z) = 2(7 - 6z)^{-4/3}$.

7. $f'(x) = \frac{3}{2}(2x + 3)^{1/2} \cdot 2 = 3(2x + 3)^{1/2}$

10. $f'(y) = -\frac{2}{3}(4 - 3y^3)^{-5/3} \cdot (-9y^2) = 6y^2(4 - 3y^3)^{-5/3}$

13. $f(x) = (2x^2 + 1)^{1/2}$, so $f'(x) = 2x(2x^2 + 1)^{-1/2}$.

16. $g(t) = (3t^5)^{-1/2}$, so $g'(t) = -\frac{1}{2}(3t^5)^{-3/2} \cdot 15t^4$.

19. $g(x) = (x - 2x^3)^{-4/3}$, so $g'(x) = -\frac{4}{3}(x - 2x^3)^{-7/3}(1 - 6x^2)$.

22. $g(x) = \left(\dfrac{2x+1}{x-1}\right)^{1/2}$, so $g'(x) = \dfrac{1}{2}\left(\dfrac{2x+1}{x-1}\right)^{-1/2} \cdot \dfrac{(x-1) \cdot 2 - (2x+1) \cdot 1}{(x-1)^2} = \dfrac{-3}{2(x-1)^2}\sqrt{\dfrac{x-1}{2x+1}}$.

25. $f'(x) = 3\left(x - \dfrac{1}{x}\right)^2 \cdot \left(1 + \dfrac{1}{x^2}\right)$

28. $h'(x) = \dfrac{5}{3}\left(\dfrac{x}{1+x^2}\right)^{2/3} \cdot \dfrac{1-x^2}{(1+x^2)^2}$

31. $f'(x) = (3 - 4x)^{1/2} - 2x(3 - 4x)^{-1/2} = 3(1 - 2x)(3 - 4x)^{-1/2}$

34. $f'(x) = -\frac{1}{2}(1-x)^{-1/2}(2-x)^{1/3} - \frac{1}{3}(1-x)^{1/2}(2-x)^{-2/3}$

37. $f'(x) = \dfrac{2(3x+4)^5 - 15(3x+4)^4(2x-1)}{(3x+4)^{10}}$, which can be simplified to $f'(x) = \dfrac{23 - 24x}{(3x+4)^6}$.

40. $f'(x) = -5(1 - 3x^4)^4(12x^3)(4 - x)^{1/3} - \frac{1}{3}(1 - 3x^4)^5(4 - x)^{-2/3}$

43. $g'(t) = \frac{1}{2}[t + (t + t^{1/2})^{1/2}]^{-1/2}[1 + \frac{1}{2}(t + t^{1/2})^{-1/2}(1 + \frac{1}{2}t^{-1/2})] = \dfrac{1 + \dfrac{1 + \dfrac{1}{2\sqrt{t}}}{2\sqrt{t + \sqrt{t}}}}{2\sqrt{t + \sqrt{t + \sqrt{t}}}}$

46. $\dfrac{dy}{dx} = \dfrac{4 - 2x^2}{\sqrt{4 - x^2}}$, so there are horizontal tangents at $(\sqrt{2}, 2)$ and $(-\sqrt{2}, -2)$. There are vertical tangents at $(2, 0)$ and $(-2, 0)$.

49. $\dfrac{dy}{dx} = \dfrac{1}{(1 - x^2)^{3/2}}$, so there are no horizontal tangents. There are no vertical tangents because $y(x)$ is undefined at $x = \pm 1$.

52. When $x = 8$, $y = f(8) = 6$ and $f'(8) = \frac{1}{4}$. The equation of the tangent line is $y - 6 = \frac{1}{4}(x - 8)$, or $4y - x = 16$.

55. When $x = 0$, $y = f(0) = 0$ and $f'(0) = 2$. The equation of the tangent line is $y = 2x$.

58. Matches (f).

61. Matches (e).

64. $dV/dS = \frac{1}{4}\sqrt{S/\pi}$, and $S = 400\pi$ when the radius of the sphere is 10, so the answer is 5 (in appropriate units).

67. The line tangent to the parabola $y = x^2$ at the point $Q(a, a^2)$ has slope $2a$, so the normal to the parabola at Q has slope $-1/(2a)$. The normal also passes through $P(18, 0)$, so we can find its slope another way—by using the two-point formula. Thus

$$-\frac{1}{2a} = \frac{a^2 - 0}{a - 18};$$

$$18 - a = 2a^3;$$

$$2a^3 + a - 18 = 0.$$

By inspection, $a = 2$ is a solution. Thus $a - 2$ is a factor of the cubic, so

$$2a^3 + a - 18 = (a - 2)(2a^2 + 4a + 9).$$

The quadratic factor has negative discriminant, so $a = 2$ is the only real solution of $2a^3 + a - 18 = 0$. Therefore the normal line has slope $-1/4$ and equation $x + 4y = 18$.

Section 3.5

1. Maximum value 2 at $x = -1$ because f is a decreasing function; no minimum value because $(1, 0)$ is not on the graph.

4. No maximum value because $\lim\limits_{x \to 0^+} f(x) = +\infty$. Minimum value 1 at $x = 1$ because f is a decreasing function.

7. The graph of f is increasing, so f has a minimum at $(-1, 0)$ and a maximum at $(1, 2)$.

10. Minimum 4 at $x = 1/2$, no maximum because $f(x) \to +\infty$ as $x \to 1^-$ and as $x \to 0^+$.

13. Because $h'(x) = -2x$ is never zero on the domain $[1, 3]$ of h, the extrema can occur only at the endpoints of the domain. And $h(3) = -5 < 3 = h(1)$, so the minimum value of h is -5 and its maximum value is 3.

16. $h'(x) = 2x + 4$; $h'(x) = 0$ when $x = -2$. $h(-3) = 4$, $h(-2) = 3$ (minimum), and $h(0) = 7$ (maximum).

19. $h'(x) = 1 - \dfrac{4}{x^2}$; $h'(x) = 0$ when $x = \pm 2$, but -2 is not in the domain of h. The minimum is $h(2) = 4$ and the maximum value of h is $h(1) = h(4) = 5$.

22. $f'(x) = 2x - 4$; $f'(x) = 0$ when $x = 2$. $f(0) = 3$ (maximum) and $f(2) = -1$ (minimum).

25. $f'(x) = 3x^2 - 6x - 9 = 3(x + 1)(x - 3)$; $f'(x) = 0$ when $x = -1$ and when $x = 3$. $f(-2) = 3$, $f(-1) = 10$ (maximum), $f(3) = -22$ (minimum), and $f(4) = -15$.

28. $f'(x) = -2$ for $1 < x < 3/2$, $f'(x) = 2$ for $3/2 < x < 2$; $f'(3/2)$ does not exist. $f(1) = 1 = f(2)$ (maximum) and $f(3/2) = 0$ (minimum).

31. $f'(x) = 150x^2 - 210x + 72 = 6(5x - 3)(5x - 4)$; $f'(x) = 0$ when $x = 3/5$ and when $x = 4/5$. $f(0) = 0$ (minimum), $f(3/5) = 16.2$, $f(4/5) = 16$, and $f(1) = 17$ (maximum).

34. $f'(x) = \dfrac{1 - x^2}{(1 + x^2)^2}$; $f'(x)$ always exists and $f'(x) = 0$ when $x = 1$ (and when $x = -1$, but the latter is not in the domain of f). $f(0) = 0$ (minimum), $f(1) = 1/2$ (maximum), and $f(3) = 0.3$.

37. $f'(x) = \dfrac{1 - 2x^2}{(1 - x^2)^{1/2}}$; $f'(x) = 0$ when $x = \pm\sqrt{2}/2$; $f'(x)$ does not exist when $x = \pm 1$ (the endpoints of the domain of f). $f(-1) = 0$, $f(-\sqrt{2}/2) = -1/2$ (minimum), $f(\sqrt{2}/2) = 1/2$ (maximum), and $f(1) = 0$.

40. $f'(x) = \dfrac{1 - 3x}{2\sqrt{x}}$; $f'(x) = 0$ when $x = 1/3$, and $f'(0)$ does not exist. $f(0) = 0$, $f(1/3) = 2\sqrt{3}/9$ (maximum), and $f(4) = -6$ (minimum).

43. $f'(x) = 0$ if x is not an integer; $f'(x)$ does not exist if x is an integer.

46. Every real number that is an integer or half an odd integer is a critical point of f; f' does not exist at each such real number.

49. (d)

52. (e)

Section 3.6

1. With $x > 0$, $y > 0$, and $x + y = 50$, we are to maximize the product $P = xy$.

$$P = P(x) = x(50 - x) = 50x - x^2, \qquad 0 < x < 50$$

($x < 50$ because $y > 0$.) The product is not maximal if we let $x = 0$ or $x = 50$, so we adjoin the endpoints to the domain of P; thus the continuous function $P(x) = 50x - x^2$ has a global maximum on the closed interval $[0, 50]$, and the maximum does *not* occur at either endpoint. Because f is differentiable, the maximum must occur at a point where $P'(x) = 0$: $50 - 2x = 0$, and so $x = 25$. Because this is the only critical point of P, it follows that $x = 25$ maximizes $P(x)$. When $x = 25$, $y = 50 - 25 = 25$, so the two positive real numbers with sum 50 and maximum possible product are 25 and 25.

4. If the side of the pen parallel to the wall has length x and the two perpendicular sides both have length y, then we are to maximize area $A = xy$ given $x + 2y = 600$. Thus

$$A = A(y) = y(600 - 2y), \qquad 0 \le y \le 300.$$

Adjoining the endpoints to the domain is allowed because the maximum we seek occurs at neither endpoint. Therefore the maximum occurs at an interior critical point. We have $A'(y) = 600 - 4y$, so the only critical point of A is $y = 150$. Then $y = 150$, we have $x = 300$, so the maximum possible area that can be enclosed is 45,000 m^2.

7. If the two numbers are x and y, then we are to minimize $S = x^2 + y^2$ given $x > 0$, $y > 0$, and $x + y = 48$. So $S(x) = x^2 + (48 - x)^2$, $0 \le x \le 48$. Here we adjoin the endpoints to the domain to ensure the existence of a maximum, but we must test the values of S at these endpoints because it is not immediately clear that neither $S(0)$ nor $S(48)$ yields the maximum value of S. Now $S'(x) = 2x - 2(48 - x)$; the only interior critical point of S is $x = 24$, and when $x = 24$, $y = 24$ as well. Now $S(0) = (48)^2 = 2304 = S(48) > 1152 = S(24)$, so the answer is 1152.

10. Draw a cross section of the cylindrical log—a circle of radius r. Inscribe in this circle a cross section of the beam—a rectangle of width w and height h. Draw a diagonal of the rectangle; the Pythagorean theorem yields $x^2 + h^2 = 4r^2$. The strength S of the beam is given by $S = kwh^2$ where k is a positive constant. Because $h^2 = 4r^2 - w^2$, we have

$$S = S(w) = kw(4r^2 - w^2) = k(4wr^2 - w^3)$$

with natural domain $0 < w < 2r$. We adjoin the endpoints to this domain; this is permissible because $S = 0$ at each, and so is not maximal. Next, $S'(w) = k(4r^2 - 3w^2)$; $S'(w) = 0$ when $3w^2 = 4r^2$, and the corresponding (positive) value of w yields the maximum of S (we know that $S(w)$ must have a maximum on $[0, 2r]$ because of the continuity of S on this interval, and we also know that the maximum does not occur at either endpoint, so there is only one possible location for the maximum). At maximum, $h^2 = 4r^2 - w^2 = 3w^2 - w^2$, so $h = w\sqrt{2}$ describes the shape of the beam of greatest strength.

13. If the rectangle has sides x and y, then $x^2 + y^2 = 16^2$ by the Pythagorean theorem. The area of the rectangle is then

$$A(x) = 2\sqrt{256 - x^2}, \qquad 0 \le x \le 16.$$

A positive quantity is maximized exactly when its square is maximized, so in place of A we maximize

$$f(x) = (A(x))^2 = 256x^2 - x^4.$$

The only solutions of $f'(x) = 0$ in the domain of A are $x = 0$ and $x = 8\sqrt{2}$; the former minimizes $A(x)$ and the latter yields its maximum value, 128.

16. Let $P(x, 0)$ be the lower right-hand corner point of the rectangle. The rectangle then has base $2x$, height $4 - x^2$, and thus area

$$A(x) = 2x(4 - x^2) = 8x - 2x^3, \qquad 0 \le x \le 2.$$

Now $A'(x) = 8 - 6x^2$; $A'(x) = 0$ when $x = 2\sqrt{3}/3$. Because $A(0) = 0$, $A(2) = 0$, and $A(2\sqrt{3}/3) \ge 0$, the maximum possible area is $32\sqrt{3}/9$.

19. Let x be the length of the edge of each of the twelve small squares. Then each of the three cross-shaped pieces will form boxes with base length $1 - 2x$ and height x, so each of the three will have volume $x(1 - 2x)^2$. Both of the two cubical boxes will have edge x and thus volume x^3. So the total volume of all five boxes will be

$$V(x) = 3x(1 - 2x)^2 + 2x^3 = 14x^3 - 12x^2 + 3x, \qquad 0 \le x \le \frac{1}{2}.$$

Now $V'(x) = 42x^2 - 24x + 3$; $V'(x) = 0$ when $14x^2 - 8x - 1 = 0$. The quadratic formula gives the two solutions $\dfrac{4 \pm \sqrt{2}}{14}$. These are approximately 0.3867 and 0.1847, and both lie in the domain of V. Now $V(0) = 0$, $V(0.1847) \approx 0.2329$, $V(0.3867) \approx 0.1752$, and $V(0.5) = 0.25$. Therefore, to maximize V, one must cut each of the three large squares into four smaller squares of side length $\frac{1}{2}$ each and form the resulting twelve squares into two cubes. At maximum volume there will be only two boxes, not five.

22. Let r be the radius of the circle and x the edge of the square. We are to maximize total area $A = \pi r^2 + x^2$ given the side condition $2\pi r + 4x = 100$. From the last equation we infer that

$$x = \frac{100 - 2\pi r}{4} = \frac{50 - \pi r}{2}.$$

So

$$A = A(r) = \pi r^2 + \frac{1}{4}(50 - \pi r)^2 = (\pi + \frac{1}{4}\pi^2)r - 25\pi r + 625$$

for $0 \le r \le 50/\pi$ (because $x \ge 0$). Now

$$A'(r) = 2(\pi + \frac{1}{4}\pi^2)r - 25\pi;$$

$$A'(r) = 0 \text{ when } r = \frac{25}{2 + \dfrac{\pi}{2}} = \frac{50}{\pi + 4};$$

that is, when $r \approx 7$. Finally, $A(0) = 625$,

$$A\left(\frac{50}{\pi}\right) \approx 795.77 \quad \text{and} \quad A\left(\frac{50}{\pi + 4}\right) \approx 350.06$$

Results: For minimum area, use a circle of radius $50/(\pi + 4) \approx 7.00124$ (cm) and a square of edge length $100/(\pi + 4)$ (cm). For maximum area, bend all the wire into a circle of radius $50/\pi$.

25. Let the dimensions of the box be x by x by y. We are to maximize $V = x^2 y$ subject to some conditions on x and y. According to the poster on the wall of the Bogart, Georgia, Post Office, the *length* of the box is the larger of x and y, and the *girth* is measured around the box in a plane perpendicular to its length.

Case 1: $x < y$. Then the length is y, the girth is $4x$, and the mailing constraint is $4x + y \le 100$. It is clear that we take $4x + y = 100$ to maximize V, so that

$$V = V(x) = x^2(100 - 4x) = 100x^2 - 4x^3, \quad 0 \le x \le 25.$$

Then $V'(x) = 4x(50 - 3x)$; $V'(x) =$ for $x = 0$ and for $x = 50/3$. But $V(0) = 0$, $V(25) = 0$, and $V(50/3) = 250{,}000/27 \approx 9259$ (in.^{3}). The latter is the maximum in Case 1.

Case 2: $x \ge y$. Then the length is x and the girth is $2x + y$, though you may get some argument from a postal worker who may insist that it's $4x$. So $3x + 2y = 100$, and thus

$$V = V(x) = x^2\left(\frac{100 - 3x}{2}\right) = 50x^2 - \frac{3}{2}x^3, \qquad 0 \le x \le 100/3.$$

Then $V'(x) = 100x - \dfrac{9}{2}x^2$; $V'(x) = 0$ when $x = 0$ and when $x = 200/9$. But $V(0) = 0$, $V(100/3) = 0$, and $V(200/9) = 2{,}000{,}000/243 \approx 8230$ (in.^{3}).

Case 3: You lose the argument in Case 2. Then the box has length x and girth $4x$, so $5x = 100$; thus $x = 20$. To maximize the total volume, no calculus is needed—let $y = x$. Then the box of maximum volume will have volume $20^3 = 8000$ (in.^{3}).

Answer: The maximum is $\dfrac{250{,}000}{27}$ in.^{3}.

28. Let x denote the number of workers hired. Each worker will pick $900/x$ bushels; each worker will spend $180/x$ hours picking beans. The supervisor cost will be $1800/x$ dollars, and the cost per worker will be $8 + 900/x$ dollars. Thus the total cost will be

$$C(x) = 8x + 900 + \frac{1800}{x}, \qquad 1 \le x.$$

It is clear that large values of x make $C(x)$ large, so the global minimum of $C(x)$ occurs either at $x = 1$ or where $C'(x) = 0$. Assume for the moment that x can take on all real number values in $[1, \infty)$, not merely integral values, so that C' is defined. Then

$$C'(x) = 8 - \frac{1800}{x^2}; \quad C'(x) = 0 \text{ when } x^2 = 225.$$

Thus $C'(15) = 0$. Now $C(1) = 2708$ and $C(15) = 1140$, so fifteen workers should be hired; the cost to pick each bushel will be approximately $1.27.

31. Let x be the number of five-cent fare increases. The resulting revenue will be

$$R(x) = (150 + 5x)(600 - 4x), \qquad -15 \le x \le 15$$

(the revenue is the product of the price and the number of passengers). Now

$$R(x) = 90000 - 3000x - 200x^2;$$

$$R'(x) = -3000 - 400x; \qquad R'(x) = 0 \text{ when } x = -7.5.$$

Because the fare must be an integral number of cents, we check $R(-7) = 1012 = R(-8)$ (dollars). Answer: The fare should be either $1.10 or $1.15; this is a reduction of 40 or 35 cents, respectively, and each results in the maximum possible revenue of $1012 per day.

34. Let the circle have equation $x^2 + y^2 = 1$ and let (x, y) denote the coordinates of the upper right-hand vertex of the trapezoid (Fig. 3.6.25). Then the area A of the trapezoid is the product of its altitude y and the average of the lengths of its two bases, so

$$A = \frac{1}{2}y(2x + 2) \text{ where } y^2 = 1 - x^2.$$

A positive quantity is maximized when its square is maximized, so we maximize instead

$$f(x) = A^2 = (x + 1)^2(1 - x^2)$$
$$= 1 + 2x - 2x^3 - x^4, \qquad 0 \le x \le 1.$$

Because $f(0) = 0 = f(1)$, f is maximized when $f'(x) = 0$:

$$0 = 2 - 6x^2 - 4x^3 = 2(1 + x)^2(1 - 2x).$$

But the only solution of $f'(x) = 0$ in the domain of f is $x = 1/2$. Finally, $f(1/2) = 27/16$, so the maximum possible area of the trapezoid is $\frac{3}{4}\sqrt{3}$. This is just over 41% of the area of the circle, so the answer meets the test of plausibility.

37. We are to maximize volume $V = \frac{1}{3}\pi r^2 h$ given $r^2 + h^2 = 100$. The latter relation enables us to write

$$V = V(h) = \frac{1}{3}\pi(100 - h^2)h = \frac{1}{3}\pi(100h - h^3), \qquad 0 \le h \le 10.$$

Now $V'(h) = \frac{1}{3}\pi(100 - 3h^2)$, so $V'(h) = 0$ when $3h^2 = 100$, thus when $h = \frac{10}{3}\sqrt{3}$. But $V(h) = 0$ at the endpoints of its domain, so the latter value of h maximizes V, and its maximum value is $\frac{2000}{27}\pi\sqrt{3}$.

40. If the base of the L has length x, then the vertical part has length $60 - x$. Place the L with its corner at the origin in the xy-plane, its base on the nonnegative x-axis, and the vertical part on the nonnegative y-axis. The two ends of the L have coordinates $(0, 60 - x)$ and $(x, 0)$, so they are at distance

$$d = d(x) = \sqrt{x^2 + (60 - x)^2}, \qquad 0 \le x \le 60.$$

A positive quantity is minimized when its square is minimal, so we minimize

$$f(x) = d^2 = x^2 + (60 - x)^2, \qquad 0 \le x \le 60.$$

Then $f'(x) = 2x - 2(60 - x) = 4x - 120$; $f'(x) = 0$ when $x = 30$. Now $f(0) = f(60) = 3600$, whereas $f(30) = 1800$. So $x = 30$ minimizes $f(x)$ and thus $d(x)$. The minimum possible distance between the two ends of the wire is therefore $d(30) = 30\sqrt{2}$.

43. Examine the plank on the right on Fig. 3.6.10. Let its height be $2y$ and its width (in the x-direction) be z. The total area of the four small rectangles in the figure is then $A = 4 \cdot z \cdot 2y = 8yz$. The circle has radius 1, and by Problem 35 the large inscribed square has dimensions $\sqrt{2}$ by $\sqrt{2}$. Thus

$$\left(\frac{1}{2}\sqrt{2} + z\right)^2 + y^2 = 1;$$

This implies that

$$y = \sqrt{\frac{1}{2} - z\sqrt{2} - z^2}\,.$$

Therefore

$$A(z) = 8z\sqrt{\frac{1}{2} - z\sqrt{2} - z^2}, \qquad 0 \le z \le 1 - \frac{1}{2}\sqrt{2}.$$

Now $A = 0$ at each endpoint of its domain, and

$$A'(z) = \frac{4\sqrt{2}\,(1 - 3z\sqrt{z} - 4z^2)}{\sqrt{1 - 2z\sqrt{2} - 2z^2}}.$$

So $A'(z) = 0$ when $z = \dfrac{-3\sqrt{2} \pm \sqrt{34}}{8}$; we discard the negative solution, and find that when $A(z)$ is maximized,

$$z = \frac{-3\sqrt{2} + \sqrt{34}}{8} \approx 0.198539,$$

$$2y = \frac{\sqrt{7 - \sqrt{17}}}{2} \approx 0.848071, \text{ and}$$

$$A(z) = \frac{\sqrt{142 + 34\sqrt{17}}}{2} \approx 0.673500.$$

The four small planks use just under 59% of the wood that remains after the large plank is cut, a very efficient use of what might be scrap lumber.

46. Set up a coordinate system in which the factory is located at the origin and the power station at (L, W) in the xy-plane—$L = 4500$, $W = 2000$. Part of the path of the power cable will be straight along the river bank and part will be a diagonal running under water. It makes no difference whether the straight part is adjacent to the factory or to the power station, so we assume the former. Thus we suppose that the power cable runs straight from $(0,0)$ to $(x,0)$, then straight from $(x,0)$ to (L,W), where $0 \leq x \leq L$. Let y be the length of the diagonal stretch of the cable. Then by the Pythagorean theorem,

$$W^2 + (L - x)^2 = y^2, \text{ so } y = \sqrt{W^2 + (L - x)^2}.$$

The cost C of the cable is $C = kx + 3ky$ where k is the cost per unit distance of over-the-ground cable. Therefore the total cost of the cable is

$$C(x) = kx + 3k\sqrt{W^2 + (L - x)^2}, \qquad 0 \leq x \leq L.$$

It will not change the solution if we assume that $k = 1$, and in this case we have

$$C'(x) = 1 - \frac{3(L - x)}{\sqrt{W^2 + (L - x)^2}}.$$

Next, $C'(x) = 0$ when $x^2 + (L - x)^2 = 9(L - x)^2$, and this leads to the solution

$$x = L - \frac{1}{4}W\sqrt{2} \text{ and } y = \frac{3}{4}W\sqrt{2}.$$

It is not difficult to verify that the latter value of x yields a value of C smaller than either $C(0)$ or $C(L)$. Answer: Lay the cable $x = 4500 - 500\sqrt{2} \approx 3793$ meters along the bank and $y = 1500\sqrt{2} \approx 2121$ meters diagonally across the river.

49. We are to minimize total cost

$$C = c_1\sqrt{a^2 + x^2} + c_2\sqrt{(L - x)^2 + b^2}.$$

$$C'(x) = \frac{c_1 x}{\sqrt{a^2 + x^2}} - \frac{c_2(L - x)}{\sqrt{(L - x)^2 + b^2}};$$

$$C'(x) = 0 \text{ when } \frac{c_1 x}{\sqrt{a^2 + x^2}} = \frac{c_2(L - x)}{\sqrt{(L - x)^2 + b^2}}.$$

The result in Part (a) is equivalent to the last equation. For Part (b), assume that $a = b = c_1 = 1$, $c_2 = 2$, and $L = 4$. Then we obtain

$$\frac{x}{\sqrt{1 + x^2}} = \frac{2(4 - x)}{\sqrt{(4 - x)^2 + 1}};$$

$$\frac{x^2}{1 + x^2} = \frac{4(16 - 8x + x^2)}{16 - 8x + x^2 + 1};$$

$$x^2(17 - 8x + x^2) = (4 + 4x^2)(16 - 8x + x^2);$$

$$17x^2 - 8x^3 + x^4 = 64 - 32x + 68x^2 - 32x^3 + 4x^4.$$

Therefore we wish to solve $f(x) = 0$ where

$$f(x) = 3x^4 - 24x^4 + 51x^2 - 32x + 64.$$

Now $f(0) = 64$, $f(1) = 62$, $f(2) = 60$, $f(3) = 22$, and $f(4) = -16$. Because $f(3) > 0 > f(4)$, we interpolate to estimate the zero of $f(x)$ between 3 and 4; it turns out that interpolation gives $x \approx 3.58$. Subsequent interpolation yields the more accurate estimate $x \approx 3.45$. (The equation $f(x) = 0$ has exactly two solutions, $x \approx 3.452462314$ and $x \approx 4.559682567$.)

52. Let the horizontal piece of wood have length $2x$ and the vertical piece have length $y + z$ where y is the length of the part above the horizontal piece and z the length of the part below it. Then

$$y = \sqrt{4 - x^2} \text{ and } z = \sqrt{16 - x^2}.$$

Also the kite area is $A = x(y + z)$; $\dfrac{dA}{dx} = 0$ implies that

$$y + z = \frac{x^2}{y} + \frac{x^2}{z}.$$

Multiply each side of the last equation by yz to obtain

$$y^2 z + yz^2 = x^2 z + x^2 y,$$

so that

$$yz(y + z) = x^2(y + z);$$
$$x^2 = yz;$$
$$x^4 = y^2 z^2 = (4 - x^2)(16 - x^2);$$
$$x^4 = 64 - 20x^2 + x^4;$$
$$20x^2 = 64;$$
$$x = \frac{4}{5}\sqrt{5}, \ y = \frac{2}{5}\sqrt{5}, \ z = \frac{8}{5}\sqrt{5}.$$

Therefore $L_1 = \frac{8}{5}\sqrt{5} \approx 3.5777$ and $L_2 = 2\sqrt{5} \approx 4.47214$ for maximum area.

55. Let V_1 and V_2 be the volume functions of problems 53 and 54, respectively.
$V_1'(x) = \dfrac{20\sqrt{5}\left(4x - x^2\right)}{3\sqrt{5} - x}$, which is zero at $x = 0$ and at $x = 4$, and $V_2'(x) = \dfrac{10\sqrt{5}\left(8x - x^2\right)}{3\sqrt{10} - x}$, which
is zero at $x = 0$ and at $x = 8$ as expected. $\dfrac{V_2(8)}{V_1(4)} = 2\sqrt{2}.$

Section 3.7

1. $f'(x) = 6\sin x \cos x$

4. $f'(x) = \dfrac{1}{2\sqrt{x}}\sin x + \sqrt{x}\cos x$

7. $f'(x) = \cos^3 x - 2\sin^2 x \cos x$

10. $g'(t) = 6(2 - \cos^2 t)^2 \sin t \cos t$

13. $f'(x) = 2\sin x + 2x \cos x - 6x \cos x + 3x^2 \sin x$

16. $f'(x) = 7\cos 5x \cos 7x - 5\sin 5x \sin 7x$

19. $g'(t) = -\dfrac{5}{2}(\cos 3t + \cos 5t)^{3/2}(3\sin 3t + 5\sin 5t)$

22. $\dfrac{dy}{dx} = \dfrac{-2x\sin 2x - \cos 2x}{x^2}$

25. $\dfrac{dy}{dx} = 2\cos 2x \cos 3x - 3\sin 2x \sin 3x$

28. $\dfrac{dy}{dx} = -\dfrac{\sin\sqrt{x}}{4\sqrt{x}\sqrt{\cos\sqrt{x}}}$

31. $\dfrac{dy}{dx} = \dfrac{\cos 2\sqrt{x}}{\sqrt{x}}$

34. $\dfrac{dy}{dx} = 2x \cos \dfrac{1}{x} + \sin \dfrac{1}{x}$

37. $\dfrac{dy}{dx} = \frac{1}{2}x^{-1/2}(x - \cos x)^3 + 3x^{1/2}(x - \cos x)^2(1 + \sin x)$

40. $\dfrac{dy}{dx} = \dfrac{(\cos x) \cos \left(1 + \sqrt{\sin x}\right)}{2\sqrt{\sin x}}$

43. $\dfrac{dx}{dt} = 7 \sec^2 t \, \tan^6 t$

46. $\dfrac{dx}{dt} = \dfrac{5t^5 \sec t^5 \tan t^5 - \sec t^5}{t^2}$

49. $\dfrac{dx}{dt} = \dfrac{2 \cot \dfrac{1}{t^2} \csc \dfrac{1}{t^2}}{t^3}$

52. $\dfrac{dx}{dt} \equiv 0$

55. $\dfrac{dx}{dt} = [\sec(\sin t) \, \tan(\sin t)] \cos t$

58. $\dfrac{dx}{dt} = \dfrac{(1 + \tan t) \sec t \tan t - \sec^3 t}{(1 + \tan t)^2} = \dfrac{\sec t \tan t - \sec t}{(1 + \tan t)^2}$

61. $\dfrac{dy}{dx} = -x \sin x + \cos x$, so the slope of the tangent at $x = \pi$ is $-\pi \sin \pi + \cos \pi = -1$. Since $y(\pi) = -\pi$, the equation of the tangent line is $y + \pi = -(x - \pi)$, or $y = -x$.

64. $\dfrac{dy}{dx} = 2 \sin \left(\dfrac{\pi x}{3}\right) \cos \left(\dfrac{\pi x}{3}\right)$, so the slope of the tangent at $x = 5$ is $2 \sin \dfrac{5\pi}{3} \cos \dfrac{5\pi}{3} = -\frac{1}{2}\sqrt{3}$. Since $y(5) = \dfrac{9}{4\pi}$, the equation of the tangent line is $y - \dfrac{9}{4\pi} = -\frac{1}{2}\sqrt{3}\,(x - 5)$, or $y = -\dfrac{x\sqrt{3}}{2} + \dfrac{9 + 10\pi\sqrt{3}}{4\pi}$.

67. $\dfrac{dy}{dx} = \cos^2 x - \sin^2 x$. This derivative is zero for $x = \frac{1}{4}\pi + n\pi$ and at $x = \frac{3}{4}\pi + n\pi$ for any integer n. The tangent line is horizontal at all points of the form $\left(n\pi + \frac{1}{4}\pi, \frac{1}{2}\right)$ and at all points of the form $\left(n\pi + \frac{3}{4}\pi, -\frac{1}{2}\right)$ where n is an integer.

70. $\dfrac{dy}{dx} = \dfrac{-13 \cos x}{(3 + \sin x)^2}$. The two lines are $y \equiv \frac{17}{4}$ and $y \equiv \frac{15}{2}$.

71. For example, if $f(x) = \sec x = \dfrac{1}{\cos x}$, then

$$f'(x) = -\dfrac{-\sin x}{\cos^2 x} = \dfrac{1}{\cos x} \cdot \dfrac{\sin x}{\cos x} = \sec x \tan x.$$

73. Write $R = R(\alpha) = \dfrac{1}{32} v^2 \sin 2\alpha$. Then

$$R'(\alpha) = \dfrac{1}{16} v^2 \cos 2\alpha,$$

which is zero when $\alpha = \pi/4$ (we assume $0 \le \alpha \le \pi/2$). Because R is zero at the endpoints of its domain, we conclude that $\alpha = \pi/4$ maximizes the range R.

76. Draw a figure in which the airplane is located at $(0, 25000)$ and the fixed point on the ground is located at $(x, 0)$. A line connecting the two produces a triangle with angle θ at $(x, 0)$. This angle is also the angle of depression of the pilot's line of sight, and when $\theta = 65°$, $d\theta/dt = 1.5°/\text{s}$. Now

$$\tan \theta = \dfrac{25000}{x}, \text{ so } x = 25000 \dfrac{\cos \theta}{\sin \theta},$$

thus

$$\frac{dx}{d\theta} = -\frac{25000}{\sin^2\theta}.$$

The speed of the airplane is

$$-\frac{dx}{dt} = \frac{25000}{\sin^2\theta} \cdot \frac{d\theta}{dt}.$$

When $\theta = \dfrac{13}{36}\pi$, $\dfrac{d\theta}{dt} = \dfrac{\pi}{120}$. So the ground speed of the airplane is

$$\frac{25000}{\sin^2\left(\dfrac{13\pi}{36}\right)} \cdot \frac{\pi}{120} \approx 796.81 \ (\text{ft/s}).$$

Answer: About 543.28 mi/h.

79. The cross section of the trough is a trapezoid with short base 2, long base $2 + 4\cos\theta$, and height $2\sin\theta$. Thus its cross-sectional area is

$$A(\theta) = \frac{2 + (2 + 4\cos\theta)}{2} \cdot 2\sin\theta$$
$$= 4(\sin\theta + \sin\theta\cos\theta), \qquad 0 \le \theta \le \pi/2$$

(the real upper bound on θ is $2\pi/3$, but the maximum value of A clearly occurs in the interval $[0, \pi/2]$).

$$A'(\theta) = 4(\cos\theta + \cos^2\theta - \sin^2\theta)$$
$$= 4(2\cos^2\theta + \cos\theta - 1)$$
$$= 4(2\cos\theta - 1)(\cos\theta + 1).$$

The only solution of $A'(\theta) = 0$ in the given domain occurs when $\cos\theta = 1/2$, so that $\theta = \pi/3$. It is easy to verify that this value of θ maximizes the function A.

82. Let L be the length of the crease. Then the right triangle of which L is the hypotenuse has sides $L\cos\theta$ and $L\sin\theta$. Now $20 = L\sin\theta + L\sin\theta\cos2\theta$, so

$$L = L(\theta) = \frac{20}{(\sin\theta)(1 + \cos2\theta)}, \qquad 0 < \theta \le \frac{\pi}{4}.$$

Next, $dL/d\theta = 0$ when

$$(\cos\theta)(1 + \cos2\theta) = (\sin\theta)(2\sin2\theta);$$
$$(\cos\theta)(2\cos^2\theta) = 4\sin^2\theta\,\cos\theta;$$

so $\cos\theta = 0$ (which is impossible given the domain of L) or

$$\cos^2\theta = 2\sin^2\theta = 2 - 2\cos^2\theta; \quad \cos^2\theta = \frac{2}{3}.$$

This implies that $\cos\theta = \sqrt6/3$ and $\sin\theta = \sqrt3/3$. Because $L \to +\infty$ as $\theta \to 0^+$, we have a minimum either at the horizontal tangent just found or at the endpoint $\theta = \pi/4$. The value of L at $\pi/4$ is $20\sqrt2 \approx 28.28$ and at the horizontal tangent we have $L = 15\sqrt3 \approx 25.98$. So the shortest crease is obtained when $\cos\theta = \sqrt6/3$; that is, for θ approximately $35°\,15'\,52''$. The bottom of the crease should be one-quarter of the way across the page from the lower left-hand corner.

85. The area in question is the area of the sector minus the area of the triangle in Fig. 3.7.19 and turns out to be

$$A = \frac{1}{2}r^2\theta - r^2\cos\frac{\theta}{2}\sin\frac{\theta}{2}$$
$$= \frac{1}{2}r^2(\theta - \sin\theta) = \frac{s^2(\theta - \sin\theta)}{2\theta^2}.$$

because $s = r\theta$. Now $\dfrac{dA}{d\theta} = \dfrac{s^2(2\sin\theta - \theta\cos\theta - \theta)}{2\theta^3}$, so $\dfrac{dA}{d\theta} = 0$ when $\theta(1 + \cos\theta) = 2\sin\theta$. Let $\theta = 2x$; note that $0 < x \le \pi$ because $0 < \theta \le 2\pi$. So the condition that $\dfrac{dA}{d\theta} = 0$ becomes

$$x = \frac{\sin\theta}{1 + \cos\theta} = \tan x.$$

But this equation has no solution in the interval $(0, \pi]$. So the only possible maximum of A must occur at an endpoint of its domain, or where x is undefined because the denominator $1 + \cos\theta$ is zero—and this occurs when $\theta = \pi$. Finally, $A(2\pi) = \dfrac{s^2}{4\pi}$ and $A(\pi) = \dfrac{s^2}{2\pi}$, so the maximum area is attained when the arc is a semicircle.

88. $f'(0) = \lim\limits_{h\to 0} \dfrac{f(0 + h) - f(0)}{h} = \lim\limits_{h\to 0} \dfrac{h^2\sin\dfrac{1}{h}}{h} = \cdots$.

Section 3.8

1. $2x - 2y\dfrac{dy}{dx} = 0$, so $\dfrac{dy}{dx} = \dfrac{x}{y}$. Also, $y = \pm\sqrt{x^2 - 1}$, so $\dfrac{dy}{dx} = \pm\dfrac{x}{\sqrt{x^2 - 1}} = \dfrac{x}{y}$.

4. $3x^2 + 3y^2\dfrac{dy}{dx} = 0$, so $\dfrac{dy}{dx} = -\dfrac{x^2}{y^2}$. $y = \sqrt[3]{1 - x^3}$, so substitution results in $\dfrac{dy}{dx} = -\dfrac{x^2}{(1 - x^3)^{2/3}}$.
Explicit differentiation yields the same answer.

7. $\frac{2}{3}x^{-1/3} + \frac{2}{3}y^{-1/3}\dfrac{dy}{dx} = 0$: $\quad \dfrac{dy}{dx} = -\left(\dfrac{y}{x}\right)^{1/3}$.

10. Given: $x^5 + y^5 = 5x^2y^2$:
$$5x^4 + 5y^4\frac{dy}{dx} = 10x^2y\frac{dy}{dx} + 10xy^2$$
$$\frac{dy}{dx} = \frac{10xy^2 - 5x^4}{5y^4 - 10x^2y}$$

13. Given: $\cos^3 x + \cos^3 y = \sin(x + y)$:
$$3\left(\cos^2 x\right)(-\sin x) + 3\left(\cos^2 y\right)(-\sin y)\frac{dy}{dx} = \cos(x + y)\left(1 + \frac{dy}{dx}\right);$$
$$\frac{dy}{dx} = -\frac{\cos(x + y) + 3\cos^2 x\sin x}{\cos(x + y) + 3\cos^2 y\sin y}.$$

16. $x\dfrac{dy}{dx} + y = 0$: $\quad \dfrac{dy}{dx} = -\dfrac{y}{x}$. At $(4, -2)$ the tangent has slope $\frac{1}{2}$ and thus equation $y + 2 = \frac{1}{2}(x - 4)$.

19. $y^2 + 2xy\dfrac{dy}{dx} + 2xy + x^2\dfrac{dy}{dx} = 0$: $\quad \dfrac{dy}{dx} = -\dfrac{2xy + y^2}{2xy + x^2}$. At $(1, -2)$ the slope is zero.

22. $2x + y + x\dfrac{dy}{dx} + 2y\dfrac{dy}{dx} = 0$: $\quad \dfrac{dy}{dx} = -\dfrac{2x + y}{x + 2y}$. At $(3, -2)$ the tangent line has slope 4 and thus equation $y + 2 = 4(x - 3)$.

25. $(2/3)x^{-1/3} + (2/3)y^{-1/3}\dfrac{dy}{dx} = 0$: $\quad \dfrac{dy}{dx} = -\dfrac{y^{1/3}}{x^{1/3}}$. At $(8, 1)$ the tangent line has slope $-\frac{1}{2}$, and thus equation $y - 1 = -\frac{1}{2}(x - 8)$, or $2y = 10 - x$.

28. $2y\dfrac{dy}{dx} = 3x^2 + 14x$: $\quad \dfrac{dy}{dx} = \dfrac{3x^2 + 14x}{2y}$. At $(-3, 6)$ the tangent line has slope $-\frac{5}{4}$ and thus equation $y - 6 = -\frac{5}{4}(x + 3)$ or $4y = 9 - 5x$.

31. Here $\dfrac{dy}{dx} = \dfrac{2 - x}{y - 2}$, so horizontal tangents can occur only if $x = 2$ and $y \ne 2$. When $x = 2$, the given equation yields $y^2 - 4y - 4 = 0$, so that $y = 2 \pm \sqrt{8}$. Thus there are two points at which the tangent line is horizontal.

34. Here, $\dfrac{dy}{dx} = \dfrac{y - 2x}{2y - x}$ and $\dfrac{dx}{dy} = \dfrac{2y - x}{y - 2x}$. For horizontal tangents, $y = 2x$, and from the original equation $x^2 - xy + y^2 = 9$; it follows upon substitution of $2x$ for y that $x^2 = 3$. So the tangent line is horizontal at $(\sqrt{3}, 2\sqrt{3})$ and at $(-\sqrt{3}, -2\sqrt{3})$. Where there are vertical tangents we must have $x = 2y$, and (as above) it turns out that $y^2 = 3$; there are vertical tangents at $(2\sqrt{3}, \sqrt{3})$ and at $(-2\sqrt{3}, -\sqrt{3})$.

37. Suppose that the pile has height $h = h(t)$ at time t (seconds) and radius $r = r(t)$ then. We are given $h = 2r$ and we know that the volume of the pile at time t is

$$V = V(t) = \frac{\pi}{3} r^2 h = \frac{2}{3} \pi r^3. \quad \text{Now } \frac{dV}{dt} = \frac{dV}{dr} \cdot \frac{dr}{dt}, \text{ so } 10 = 2\pi r^2 \frac{dr}{dt}.$$

When $h = 5$, $r = 2.5$; at that time $\dfrac{dr}{dt} = \dfrac{10}{2\pi(2.5)^2} = \dfrac{4}{5\pi} \approx 0.25645$ (ft/s).

40. Let x be the distance from the ostrich to the street light and u the distance from the base of the light pole to the tip of the ostrich's shadow. Draw a figure and so label it; by similar triangles you find that $\dfrac{u}{10} = \dfrac{u - x}{5}$, and it follows that $u = 2x$. We are to find du/dt and $D_t(u - x) = du/dt - dx/dt$. But $u = 2x$, so

$$\frac{du}{dt} = 2\frac{dx}{dt} = (2)(-4) = -8; \qquad \frac{du}{dt} - \frac{dx}{dt} = -8 - (-4) = -4.$$

Answers: (a) $+8$ ft/s; (b) $+4$ ft/s.

43. Let r denote the radius of the balloon and V its volume at time t (in seconds). Then

$$V = \frac{4}{3}\pi r^3, \text{ so } \frac{dV}{dt} = 4\pi r^2 \frac{dr}{dt}.$$

We are to find dr/dt when $r = 10$, and we are given the information that $dV/dt = 100\pi$. Therefore

$$100\pi = 4\pi(10)^2 \frac{dr}{dt}\bigg|_{r=10},$$

and so at the time in question the radius is increasing at the rate of $dr/dt = \frac{1}{4} = 0.25$ (cm/s).

46. Locate the observer at the origin and the balloon in the first quadrant at $(300, y)$, where $y = y(t)$ is the balloon's altitude at time t. Let θ be the angle of elevation of the balloon (in radians) from the observer's point of view. Then $\tan \theta = y/300$. We are given $d\theta/dt = \pi/180$ rad/s. Hence we are to find dy/dt when $\theta = \pi/4$. But $y = 300 \tan \theta$, so

$$\frac{dy}{dt} = (300 \sec^2 \theta) \frac{d\theta}{dt}.$$

Substitution of the given values of θ and $d\theta/dt$ yields the answer

$$\frac{dy}{dt}\bigg|_{\theta=45°} = 300 \cdot 2 \cdot \frac{\pi}{180} = \frac{10\pi}{3} \approx 10.472 \text{ (ft/s)}.$$

49. We use $a = 10$ in the formula given in Problem 42. Then

$$V = \frac{1}{3}\pi y^2 (30 - y).$$

Hence $(-100)(0.1337) = \dfrac{dV}{dt} = \pi(20y - y^2)\dfrac{dy}{dt}$. Thus $\dfrac{dy}{dt} = -\dfrac{13 \cdot 37}{\pi y(20 - y)}$. Substitution of $y = 7$ and $y = 3$ now yields the two answers.

52. Let x be the length of the base of the rectangle and y its height. We are given $dx/dt = +4$ and $dy/dt = -3$, with units in centimeters and seconds. The area of the rectangle is $A = xy$, so

$$\frac{dA}{dt} = x\frac{dy}{dt} + y\frac{dx}{dt} = -3x + 4y.$$

Therefore when $x = 20$ and $y = 12$, we have $dA/dt = -12$, so the area of the rectangle is decreasing at the rate of 12 cm^2/s then.

55. Locate the radar station at the origin and the rocket at $(4, y)$ in the first quadrant at time t, with y in miles and t in hours. The distance z between the station and the rocket satisfies the equation $y^2 + 16 = z^2$, so $2y\dfrac{dy}{dt} = 2z\dfrac{dz}{dt}$. When $z = 5$, we have $y = 3$, and because $dz/dt = 3600$ it follows that $dy/dt = 6000$ mi/h.

58. Let x be the distance between the *Pinta* and the island at time t and y the distance between the *Niña* and the island then. We know that $x^2 + y^2 = z^2$ where $z = z(t)$ is the distance between the two ships, so

$$2z\frac{dz}{dt} = 2x\frac{dx}{dt} + 2y\frac{dy}{dt}.$$

When $x = 30$ and $y = 40$, $z = 50$. It follows from the last equation that $dz/dt = -25$ then. Answer: They are drawing closer at 25 mi/h then.

61. Let x be the radius of the water surface at time t and y the height of the water remaining at time t. If Q is the amount of water remaining in the tank at time t, then (because the water forms a cone) $Q = Q(t) = \frac{1}{3}\pi x^2 y$. But by similar triangles, $\dfrac{x}{y} = \dfrac{3}{5}$, so $x = \dfrac{3y}{5}$. So

$$Q(t) = \frac{1}{3}\pi\frac{9}{25}y^3 = \frac{3}{25}\pi y^3.$$

We are given $dQ/dt = -2$ when $y = 3$. This implies that when $y = 3$, $-2 = \dfrac{dQ}{dt} = \dfrac{9}{25}\pi y^2\dfrac{dy}{dt}$. So at the time in question,

$$\left.\frac{dy}{dt}\right|_{y=3} = -\frac{50}{81\pi} \approx -0.1965 \text{ (ft/s)}.$$

64. Let x denote the distance between the ship and A, y the distance between the ship and B, h the perpendicular distance from the position of the ship to the line AB, u the distance from A to the foot of this perpendicular, and v the distance from B to the foot of the perpendicular. At the time in question, we know that $x = 10.4$, $dx/dt = 19.2$, $y = 5$, and $dy/dt = -0.6$. From the right triangles involved, we see that $u^2 + h^2 = x^2$ and $(12.6 - u)^2 + h^2 = y^2$. Therefore

$$x^2 - u^2 = y^2 - (12.6 - u)^2. \tag{$*$}$$

We take $x = 10.4$ and $y = 5$ in Eq. ($*$); it follows that $u = 9.6$ and that $v = 12.6 - u = 3$. From Eq. ($*$), we know that

$$x\frac{dx}{dt} - u\frac{du}{dt} = y\frac{dy}{dt} + (12.6 - u)\frac{du}{dt},$$

so

$$\frac{du}{dt} = \frac{1}{12.6}\left(x\frac{dx}{dt} - y\frac{dy}{dt}\right).$$

From the data given, $du/dt \approx 16.0857$. Also, because $h = \sqrt{x^2 - u^2}$, $h = 4$ when $x = 10.4$ and $y = 9.6$. Moreover, $h\dfrac{dh}{dt} = x\dfrac{dx}{dt} - u\dfrac{du}{dt}$, and therefore

$$\left.4\frac{dh}{dt}\right|_{h=4} \approx (10.4)(19.2) - (9.6)(16.0857) \approx 11.3143.$$

Finally, $\dfrac{dh/dt}{du/dt} \approx 0.7034$, so the ship is sailing a course about $35°7'$ north *or* south of east at a speed of $\sqrt{(du/dt)^2 + (dh/dt)^2} \approx 19.67$ mi/h. It is located 9.6 miles east and 4 miles north *or* south of A, or 10.4 miles from A at a bearing of either $67°22'\,48''$ or $112°37'\,12''$.

67. Place the pole at the origin in the plane, and let the horizontal strip $0 \le y \le 30$ represent the road. Suppose that the person is located at $(x, 30)$ with $x > 0$ and is walking to the right, so $dx/dt = +5$. Then the distance from the pole to the person will be $\sqrt{x^2 + 900}$. Let z be the length of the person's shadow. By similar triangles it follows that $2z = \sqrt{x^2 + 900}$, so $4z^2 = x^2 + 900$, and thus $8z\dfrac{dz}{dt} = 2x\dfrac{dx}{dt}$. When $x = 40$, we find that $z = 25$, and therefore that

$$100\dfrac{dz}{dt}\bigg|_{z=25} = 40 \cdot 5 = 200.$$

Therefore the person's shadow is lengthening at 2 ft/s at the time in question.

Section 3.9

Note: In this section your results may differ from the answers in the last one or two decimal places because of differences in calculators or in methods of solving the equations.

Note: In Problems 1 through 20, we obtained our initial estimate x_0 of the solution by linear interpolation: Write the equation of the straight line that joins $(a, f(a))$ with $(b, f(b))$ and let x_0 be the x-coordinate of the point where this line crosses the x-axis.

1. $x_0 = 2.2$; we use $f(x) = x^2 - 5$.
    Then $x_1 = 2.236363636$, $x_2 = 2.236067997$, and $x_3 = x_4 = 2.236067978$.

4. Let $f(x) = x^{3/2} - 10$. Then $x_0 = 4.628863603$. From the iterative formula

    $$x \longleftarrow x - \dfrac{x^{3/2} - 10}{\frac{3}{2}x^{1/2}},$$

    we obtain $x_1 = 4.641597575$, $x_2 = 4.641588834 = x_3$.

7. $x_0 = -0.5$;
    $x_1 = -0.8108695652$, $x_2 = -0.7449619516$, $x_3 = -0.7402438226$, $x_4 = -0.7402217821 = x_5$

10. Let $f(x) = x^2 - \sin x$. Then $f'(x) = 2x - \cos x$. The interpolation formula at the beginning of this solution section yields $x_0 = 0.7956861008$, and the iterative formula

    $$x \longleftarrow x - \dfrac{x^2 - \sin x}{2x - \cos x}$$

    (with calculator in *radian* mode) yields these results:
    $x_1 = 0.8867915207$, $x_2 = 0.8768492470$, $x_3 = 0.8767262342$, and $x_4 = 0.8767262154 = x_5$.

13. With $x_0 = 2.188405797$ and the iterative formula

    $$x \longleftarrow x - \dfrac{x^4(x+1) - 100}{x^3(5x + 4)},$$

    we obtain $x_1 = 2.360000254$, $x_2 = 2.339638357$, $x_3 = 2.339301099$, and $x_4 = 2.339301008 = x_5$.

16. Because $7\pi/2 \approx 10.9956$ and $4\pi \approx 12.5663$ are the nearest discontinuities of $f(x) = x - \tan x$, this function has the intermediate value property on the interval $[11, 12]$. Because $f(11) \approx -214.95$ and $f(12) \approx 11.364$, the equation $f(x) = 0$ has a solution in $[11, 12]$. We obtain $x_0 = 11.94978618$ by interpolation, and the iteration

    $$x \longleftarrow x - \dfrac{x + \tan x}{1 + \sec^2 x}$$

of Newton's method yields the successive approximations

$$x_1 = 7.457596948, \quad x_2 = 6.180210620, \quad x_3 = 3.157913273, \quad x_4 = 1.571006986;$$

after many more iterations we arrive at the answer 2.028757838 of Problem 15. The difficulty is caused by the fact that $f(x)$ is generally a very large number, so the iteration of Newton's method tends to alter the value of x excessively. A little experimentation yields the fact that $f(11.08) \approx -0.736577$ and $f(11.09) \approx 0.531158$. We begin anew on the better interval $[11.08, 11.09]$ and obtain $x_0 = 11.08581018$, $x_1 = 11.08553759$, $x_2 = 11.08553841$, and $x_3 = x_2$.

19. $x_0 = 1.538461538$, $\qquad x_1 = 1.932798309$, $\qquad x_2 = 1.819962709$,
$x_3 = 1.802569044$, $\qquad x_4 = 1.802191039$, $\qquad x_5 = 1.802190864$,
$x_6 = 1.802190864 = x_5$.
The convergence is slow because $|f'(x)|$ is large when x is near 1.8.

22. (b) With $x_0 = 1.5$ we obtain the successive approximations 1.610123, 1.586600, 1.584901, 1.584893, and 1.584893.

25. Using the first formula, we obtain $x_0 = 0.5$, $x_1 = -1$, $x_2 = 2$, $x_3 = 2.75$, $x_4 = 2.867768595$, ..., $x_{12} = 2.879385242 = x_{13}$. Wrong root! At least the method converged. With the second formula, we find $x_0 = 0.5$, error message on x_1. That's because our pocket computer won't compute the cube root of a negative number. With the alternative formula $x = \sqrt[3]{3x^2 - 1}$, we obtain $x_0 = 0.5$, $x_1 = 0.629960525$, $x_2 = 0.5754446861$, $x_3 = 0.1874852347$, $x_4 = 0.9635358111$, $x_5 = 1.213098128$, ..., $x_{62} = 2.879385240 = x_{63}$. Wrong root again. Actually, we're lucky to get a correct root in any case, because using absolute values in the formula changes it so much that we might be solving the wrong equation!

28. 0.8241323123 and −0.8241323123

34. The graphs of $y = x$ and $y = \tan x$ show that the smallest positive solution of $f(x) = x - \tan x = 0$ is between π and $3\pi/2$. With initial guess $x = 4.5$ we obtain 4.493613903, 4.493409655, 4.493409458, and 4.493409458.

37. With $x_0 = 0.25$, Newton's method yields the sequence 0.2259259259, 0.2260737086, 0.2260737138, 0.2260717138. To four places, $x = 0.2261$.

40. Plot the graph from −1 to 12 to see that this equation has exactly one solution, and that this solution lies between $x = 10$ and $x = 12$. The solution is approximately 10.9901597, which is 10.9902 rounded to four decimal places.

Chapter 3 Miscellaneous

1. $\dfrac{dy}{dx} = 2x - \dfrac{6}{x^3}$

4. $\dfrac{dy}{dx} = \frac{5}{2}(x^2 + 4x)^{3/2}(2x + 4) = 5(x + 2)(x^2 + 4x)^{3/2}$

7. $\dfrac{dy}{dx} = 4(3x - \frac{1}{2}x^{-2})^3(3 + x^{-3})$

10. $y = (5x^6)^{-1/2}$: $\dfrac{dy}{dx} = -\frac{1}{2}(5x^6)^{-3/2}(30x^5) = -\dfrac{3}{x\sqrt{5x^6}} = -\dfrac{3\sqrt{5}}{5x^4}$

13. $\dfrac{dy}{dx} = \dfrac{dy}{du} \cdot \dfrac{du}{dx} = \dfrac{-2u}{(1 + u^2)^2} \cdot \dfrac{-2x}{(1 + x^2)^2}$. Now $1 + u^2 = 1 + \dfrac{1}{(1 + x^2)^2} = \dfrac{x^4 + 2x^2 + 2}{(1 + x^2)^2}$.

So $\dfrac{dy}{du} = \dfrac{-2u}{(1 + u^2)^2} = \dfrac{-2}{1 + x^2} \cdot \dfrac{(1 + x^2)^4}{(x^4 + 2x^2 + 2)^2} = \dfrac{-2(1 + x^2)^3}{(x^4 + 2x^2 + 2)^2}$.

Therefore $\dfrac{dy}{dx} = \dfrac{-2(1 + x^2)^3}{(x^4 + 2x^2 + 2)^2} \cdot \dfrac{-2x}{(1 + x^2)^2} = \dfrac{4x(1 + x^2)}{(x^4 + 2x^2 + 2)^2}$.

16. $\dfrac{dy}{dx} = \dfrac{15x^4 - 8x}{2(3x^5 - 4x^2)^{1/2}}$

19. $2x^2 y \dfrac{dy}{dx} + 2xy^2 = 1 + \dfrac{dy}{dx}$, so $\dfrac{dy}{dx} = \dfrac{1 - 2xy^2}{2x^2 y - 1}$.

22. $\dfrac{dy}{dx} = \dfrac{(x^2 + \cos x)(1 + \cos x) - (x + \sin x)(2x - \sin x)}{(x^2 + \cos x)^2} = \dfrac{1 - x^2 - x \sin x + \cos x + x^2 \cos x}{(x^2 + \cos x)^2}$

28. $\dfrac{dy}{dx} = \dfrac{3(1 + \sqrt{x})^2}{2\sqrt{x}}(1 - 2\sqrt[3]{x})^4 + 4(1 - 2\sqrt[3]{x})^3 \left(-\tfrac{2}{3}x^{-2/3}\right)(1 + \sqrt{x})^3$

31. $\dfrac{dy}{dx} = (\sin^3 2x)(2)(\cos 3x)(-\sin 3x)(3) + (\cos^2 3x)(3 \sin^2 2x)(2 \cos 2x)$

$\phantom{\dfrac{dy}{dx}} = 6(\cos 3x \, \sin^2 2x)(\cos 3x \, \cos 2x - \sin 2x \, \sin 3x)$

$\phantom{\dfrac{dy}{dx}} = 6 \cos 3x \, \cos 5x \, \sin^2 2x$

34. $(x + y)^3 = (x - y)^2$, so $3(x + y)^2 \left(1 + \dfrac{dy}{dx}\right) = 2(x - y)\left(1 - \dfrac{dy}{dx}\right)$;

thus $\dfrac{dy}{dx} = \dfrac{2(x - y) - 3(x + y)^2}{3(x + y)^2 + 2(x - y)} = \dfrac{2(x + y)^{3/2} - 3(x + y)^2}{3(x + y)^2 + 2(x + y)^{3/2}} = \dfrac{2 - 3\sqrt{x + y}}{2 + 3\sqrt{x + y}}$.

37. $1 = (2 \cos 2y)\dfrac{dy}{dx}$, so $\dfrac{dy}{dx} = \dfrac{1}{2 \cos 2y}$. Because $\dfrac{dy}{dx}$ is undefined at $(1, \pi/4)$, there may well be a vertical tangent at that point; sure enough, $\dfrac{dx}{dy} = 0$ at $(1, \pi/4)$. So an equation of the tangent line is $x = 1$.

40. $V(x) = \tfrac{1}{3}\pi(36x^2 - x^3)$: $V'(x) = \pi x(24 - x)$. Now $\dfrac{dV}{dt} = \dfrac{dV}{dx} \cdot \dfrac{dx}{dt}$; when $x = 6$, $36\pi = -108\pi \dfrac{dx}{dt}$, so $dx/dt = -\tfrac{1}{3}$ (in./s) when $x = 6$.

43. $x \cot 3x = \dfrac{1}{3} \cdot \dfrac{3x}{\sin 3x} \to \dfrac{1}{3} \cdot 1 \cdot 1 = \dfrac{1}{3}$ as $x \to 0$.

46. $-1 \le \sin u \le 1$ for all u. So

$$-x^2 \le x^2 \sin \dfrac{1}{x^2} \le x^2$$

for all $x \ne 0$. But $x^2 \to 0$ as $x \to 0$, so the limit of the expression caught in the squeeze is also zero.

49. $g(x) = x^{-1/2}$, $f(x) = x^2 + 25$.

52. $g(x) = x^{10}$, $f(x) = \dfrac{x + 1}{x - 1}$; $h'(x) = -\dfrac{20(x + 1)^9}{(x - 1)^{11}}$.

55. $\dfrac{dV}{dA} \cdot \dfrac{dA}{dr} = \dfrac{dV}{dr}$. Now $V = \dfrac{4}{3}\pi r^3$ and $A = 4\pi r^2$, so $\dfrac{dV}{dA} \cdot 8\pi r = 4\pi r^2$, and therefore $\dfrac{dV}{dA} = \dfrac{r}{2} = \dfrac{1}{4}\sqrt{\dfrac{A}{\pi}}$.

58. Current production per well: 200 (bbl/day). Number of new wells: x ($x \ge 0$). Production per well: $200 - 5x$. Total production:

$$T = T(x) = (20 + x)(200 - 5x), \qquad 0 \le x \le 40.$$

Now $T(x) = 4000 + 100x - 5x^2$, so $T'(x) = 100 - 10x$. $T'(x) = 0$ when $x = 10$. $T(0) = 4000$, $T(40) = 0$, and $T(10) = 4500$. So $x = 10$ maximizes $T(x)$. Answer: Ten new wells should be drilled, thereby increasing total production from 4000 bbl/day to 4500 bbl/day.

61. Let one sphere have radius r; the other, s. We seek the extrema of $A = 4\pi(r^2 + x^2)$ given $\tfrac{4}{3}\pi(r^3 + s^3) = V$, a constant. We illustrate here the **method of auxiliary variables**:

$$\dfrac{dA}{dr} = 4\pi \left(2r + 2s\dfrac{ds}{dr}\right);$$

the condition $dA/dr = 0$ yields $ds/dr = -r/s$. But we also know that $\tfrac{4}{3}\pi(r^3 + x^3) = V$; differentiation of both sides of this *identity* with respect to r yields

$$\frac{4}{3}\pi\left(3r^2 + 3s^2\frac{ds}{dr}\right) = 0, \text{ and so}$$

$$3r^2 + 3s^2\left(-\frac{r}{s}\right) = 0;$$

$$r^2 - rs = 0.$$

Therefore $r = 0$ or $r = s$. Also, ds/dr is undefined when $s = 0$. So we test these three critical points. If $r = 0$ or if $s = 0$, there is only one sphere, with radius $(3V/4\pi)^{1/3}$ and surface area $(36\pi V^2)^{1/3}$. If $r = s$, then there are two spheres of equal size, both with radius $\frac{1}{2}(3V/\pi)^{1/3}$ and surface area $(72\pi V^2)^{1/3}$. Therefore, for maximum surface area, make two equal spheres. For minimum surface area, make only one sphere.

64. Let x denote the length of the two sides of the corral that are perpendicular to the wall. There are two cases to consider.

Case 1: Part of the wall is used. Let y be the length of the side of the corral parallel to the wall. Then $y = 400 - 2x$, and we are to maximize the area

$$A = xy = x(400 - 2x), \qquad 150 \le x \le 200.$$

Then $A'(x) = 400 - 4x$; $A'(x) = 0$ when $x = 100$, but that value of x is not in the domain of A. Note that $A(150) = 15000$ and that $A(200) = 0$.

Case 2: All of the wall is used. Let y be the length of fence added to one end of the wall, so that the side parallel to the wall has length $100 + y$. Then $100 + 2y + 2x = 400$, so $y = 150 - x$. We are to maximize the area

$$A = x(100 + y) = x(250 - x), \qquad 0 \le x \le 150.$$

In this case $A'(x) = 0$ when $x = 125$. And in this case $A(150) = 15000$, $A(0) = 0$, and $A(125) = 15625$.

Answer: The maximum area is 15,625 ft^2; to attain it, use all the existing wall and build a square corral.

67. Let x be the width of the base of the box, so that the base has length $2x$; let y be the height of the box. Then the volume of the box is $V = 2x^2y$, and for its total surface area to be 54 ft^2, we require $2x^2 + 6xy = 54$. Therefore the volume of the box is given by

$$V = V(x) = 2x^2\left(\frac{27 - x^2}{3x}\right) = \frac{2}{3}(27x - x^3), \qquad 0 < x \le 3\sqrt{3}.$$

Now $V'(x) = 0$ when $x^2 = 9$, so that $x = 3$. Also $V(0) = 0$, so even though $x = 0$ is not in the domain of V, the continuity of V implies that $V(x)$ is near zero for x near zero. Finally, $V\left(3\sqrt{3}\right) = 0$, so $V(3) = 36$ (ft^3) is the maximum possible volume of the box.

70. The square of the length of PQ is a function of x, $G(x) = (x - x_0)^2 + (y - y_0)^2$, which we are to maximize given the constraint $C(x) = y - f(x) = 0$. Now

$$\frac{dG}{dx} = 2(x - x_0) + 2(y - y_0)\frac{dy}{dx} \text{ and } \frac{dC}{dx} = \frac{dy}{dx} - f'(x).$$

When both vanish, $f'(x) = \dfrac{dy}{dx} = -\dfrac{x - x_0}{y - y_0}$. The line containing P and Q has slope

$$\frac{y - y_0}{x - x_0} = -\frac{1}{f'(x)},$$

and therefore this line is normal to the graph at Q.

73. As the diagram to the right suggests, we are to minimize the sum of the lengths of the two diagonals. Fermat's principle of least time may be used here, so we know that the angles at which the roads meet the shore are equal, and thus so are the tangents of those angles: $\frac{x}{1} = \frac{6-x}{2}$. It follows that the pier should be built two miles from the point on the shore nearest the first town. A short computation is sufficient to show that this actually yields the global minimum.

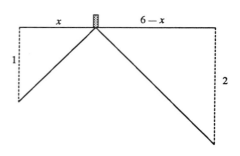

76. Here we have

$$R = R(\theta) = \frac{v^2\sqrt{2}}{16}(\cos\theta\sin\theta - \cos^2\theta) \text{ for } \pi/4 \le \theta \le \pi/2.$$

Now

$$R'(\theta) = \frac{v^2\sqrt{2}}{16}(\cos^2\theta - \sin^2\theta + 2\sin\theta\cos\theta);$$

$R'(\theta) = 0$ when $\cos 2\theta + \sin 2\theta = 0$, so that $\tan 2\theta = -1$. It follows that $\theta = 3\pi/8$ (67.5°). This yields the maximum range because $R(\pi/4) = 0 = R(\pi/2)$.

79. Use the iteration $x_{n+1} = x_n - \frac{(x_n)^5 - 75}{5(x_n)^4}$.

Results: $x_0 = 2.20379147$, $x_1 = 2.39896311$, $x_2 = 2.37206492$, $x_3 = 2.37144094$, $x_4 = 2.37144061$. Answer: 2.3714.

82. Linear interpolation gives $x_0 = -1/3$. We iterate using the formula

$$x_{n+1} = x_n - \frac{(x_n)^3 - 4x_n - 1}{3(x_n)^2 - 4}$$

to obtain the sequence -0.2525252525, -0.2541011930, -0.2541016884 of improving approximations. Answer: -0.2541.

85. Linear interpolation gives $x_0 = -0.5854549279$. If you got -0.9996955062 instead, it's because your calculator was set in degree mode. Change to radians and begin anew. Note also that a quick glance at the graphs of $y = -x$ and $y = \cos x$ shows that the equation $x + \cos x = 0$ has exactly one solution in the interval $[-2, 0]$. We use the Newton's method formula: We iterate

$$x \longleftarrow x - \frac{x + \cos x}{1 - \sin x}$$

and obtain -0.7451929664, -0.7390933178, -0.7390851332, -0.7390851332. Answer: -0.7391

88. Let $f(x) = 5(x + 1) - \cos x$. Then $f'(x) = 5 + \sin x$, and we iterate using the Newton's method formula

$$x \longleftarrow x - \frac{f(x)}{f'(x)}.$$

Linear interpolation yields the initial estimate -0.8809986055, and the succeeding approximations are -0.8712142932, -0.8712215145, and -0.8712215145. Answer: -0.8712.

91. Let $f(x) = x^5 - 3x^3 + x^2 - 23x + 19$. Then $f(-3) = -65$, $f(0) = 19$, $f(1) = 5$, and $f(3) = 121$. So there are at least three, and at most five, real solutions. Newton's method produces three real solutions, specifically $r_1 = -2.722493355$, $r_2 = 0.8012614801$, and $r_3 = 2.309976541$. If one divides the polynomial $f(x)$ by $(x - r_1)(x - r_2)(x - r_3)$, one obtains the quotient polynomial $x^2 + (0.38874466)x + 3.770552031$, which has no real roots—the quadratic formula yields the two complex roots $-0.194372333 \pm (1.932038153)i$. Consequently we have found all three real solutions.

94. We factor: $z^{3/2} - x^{3/2} = (z^{1/2})^3 - (x^{1/2})^3$
$$= (z^{1/2} - x^{1/2})(z + z^{1/2}x^{1/2} + x)$$
and $z - x = (z^{1/2})^2 - (x^{1/2})^2 = (z^{1/2} - x^{1/2})(z^{1/2} + x^{1/2})$. Therefore

$$\frac{z^{3/2} - x^{3/2}}{z - x} = \frac{z + z^{1/2}x^{1/2} + x}{z^{1/2} + x^{1/2}} \rightarrow \frac{3x}{2x^{1/2}} = \frac{3}{2}x^{1/2} \text{ as } z \rightarrow x.$$

97. The balloon has volume $V = \frac{4}{3}\pi r^3$ and surface area $A = 4\pi r^2$ where r is its radius and V, A, and r are all functions of time t. We are given $dV/dt = +10$, and we are to find dA/dt when $r = 5$.

$$\frac{dV}{dt} = 4\pi r^2 \frac{dr}{dt}, \text{ so } 10 = 4\pi \cdot 25 \cdot \frac{dr}{dt}.$$
$$\text{Thus } \frac{dr}{dt} = \frac{10}{100\pi} = \frac{1}{10\pi}.$$
$$\text{Also } \frac{dA}{dt} = 8\pi r \frac{dr}{dt}, \text{ and therefore}$$
$$\left.\frac{dA}{dt}\right|_{r=5} = 8\pi \cdot 4 \cdot \frac{1}{10\pi} = 4.$$

Answer: At 4 in.^{2}/s.

100. Let x denote the distance from plane A to the airport, y the distance from plane B to the airport, and z the distance between the two aircraft. Then

$$z^2 = x^2 + y^2 + (3 - 2)^2 = x^2 + y^2 + 1$$

and $dx/dt = -500$. Now

$$2z\frac{dz}{dt} = 2x\frac{dx}{dt} + 2y\frac{dy}{dt},$$

and when $x = 2$, $y = 2$. Therefore $z = 3$ at that time. Therefore,

$$3 \cdot (-600) = 2 \cdot (-500) + 2 \cdot \left.\frac{dy}{dt}\right|_{x=2},$$

and thus $\left.\dfrac{dy}{dt}\right|_{x=2} = -400$. Answer: Its speed is 400 mi/h.

103. The straight line through $P(x_0, y_0)$ and $Q(a, a^2)$ has slope $\dfrac{a^2 - y_0}{a - x_0} = 2a$, a consequence of the two-point formula for slope and the fact that the line is tangent to the parabola at Q. Hence $a^2 - 2ax_0 + y_0 = 0$. Think of this as a quadratic equation in the unknown a. It has two real solutions when the discriminant is positive: $(x_0)^2 - y_0 > 0$, and this establishes the conclusion in part (b). There are no real solutions when $(x_0)^2 - y_0 < 0$, and this establishes the conclusion in part (c). What if $(x_0)^2 - y_0 = 0$?

This seems a good spot to mention Descartes' Rule of Signs. In a sequence of nonzero numbers, such as

$$-3, \quad -5, \quad * \quad 2, \quad 4, \quad * \quad -1, \quad * \quad 6, \quad 6, \quad 6, \quad * \quad -2, \quad -2, \quad * \quad 12,$$

there are five *sign changes,* marked with asterisks. If some terms of such a sequence are zero, they are simply disregarded in counting the number of sign changes; thus the sequence

$$2, \quad 0, \quad 0, \quad * \quad -3, \quad -4, \quad 0, \quad 0, \quad 0, \quad * \quad 2, \quad 3, \quad * \quad -4, \quad 0,$$

has three sign changes.

Suppose that $p(x)$ is a polynomial with real coefficients, so that the polynomial equation $p(x) = 0$ has the form

$$a_0 x^n + a_1 x^{n-1} + \cdots + a_{n-1} x + a_n = 0.$$

Descartes' Rule of Signs is a classical theorem that states that the number of positive real roots of this equation is never greater than the number of sign changes in the sequence

$$a_0, \ a_1, \ a_2, \ \cdots, \ a_{n-1}, \ a_n,$$

and if less, then is less by an even number. For example, the equation

$$f(x) = x^5 + 3x^3 + x^2 - 23x + 19 = 0$$

of Problem 91 has the sequence 1, 0, -3, -23, 19 of coefficients, which has four sign changes. We already know that there are at least two and at most four positive solutions of the equation, and therefore there are either two or four. Now we consider the related equation

$$f(-x) = (-x)^5 - 3(-x)^3 + (-x)^2 - 23(-x) + 19 = 0,$$

which has the same number of positive roots as the original equation has negative roots. The sequence of coefficients in this equation is -1, 0, 3, 1, 23, 19; there is one sign change, so the original equation has exactly one negative root (because we already know of the existence of at least one).

Chapter 4: Additional Applications of the Derivative

Section 4.2

4. $dy = -\left(\dfrac{1}{(x - x^{1/2})^2}\right) \cdot (1 - \frac{1}{2}x^{-1/2})\, dx.$

10. $dy = (x^2 \cos x + 2x \sin x)\, dx.$

16. $dy = -3(1 + \cos 2x)^{1/2}\, (\sin 2x)\, dx$

19. $f(x) \approx f(0) + f'(0)(x - 0) = 1 + 2x$

22. $f'(x) = -2(1 + 3x)^{-5/3}$, so $f(x) \approx f(0) + f'(0)(x - 0) = 1 - 2x$

25. Let $f(x) = x^{1/3}$. Then $f'(x) = \frac{1}{3}x^{-2/3}$. Take $a = 27$ and $x = 25$. Then $f(x) \approx f(a) + f'(a)(x - a)$, so

$$25^{1/3} = f(25)$$
$$\approx f(27) + f'(27)(25 - 27)$$
$$= 3 + \tfrac{1}{27}(-2) \approx 2.926.$$

28. Let $f(x) = x^{1/2}$; then $f'(x) = \frac{1}{2}x^{-1/2}$. Let $a = 81$ and let $x = 80$. Then

$$\sqrt{80} = f(x) \approx f(a) + f'(a)(x - a)$$
$$= \sqrt{81} + \frac{1}{2\sqrt{81}}(-1)$$
$$= \tfrac{161}{18} \approx 8.944.$$

34. $\dfrac{1}{2} - \dfrac{\pi}{180}\sqrt{3} \approx 0.46977.$

40. If C is the circumference of the circle and r its radius, then $C = 2\pi r$. Thus $dC = 2\pi\, dr$, and so $\Delta C \approx 2\pi \Delta r$. With $r = 10$ and $\Delta r = 0.5$, we obtain $\Delta C \approx 2\pi(0.5) = \pi \approx 3.1416$. This happens to be the exact value as well (because C is a linear function of r).

46. Because v is constant, R is a function of the angle of inclination θ alone, and hence

$$dR = \frac{1}{8}v^2(\cos 2\theta)\, d\theta.$$

With $\theta = \pi/4$, $d\theta = \pi/180$ (1°), and $v = 80$, we obtain

$$\Delta R \approx \frac{1}{8}(6400)(0)\frac{\pi}{180} = 0.$$

The true value of ΔR is approximately -0.2437.

52. With the notation of the preceding solution, we now require that

$$\frac{|dS|}{S} \leq 0.0001;$$

thus, at least approximately,

$$\frac{|4\pi r\, dr|}{|2\pi r^2|} \leq 0.0001.$$

Hence

$$2\left|\frac{dr}{r}\right| \leq 0.0001,$$

which implies that $|dr|/r \leq 0.00005$. Answer: With percentage error not exceeding 0.005%.

55. We plotted $f(x) = 1/x$ and its linear approximation $L(x) = \frac{1}{2} + \frac{1}{4}(2 - x)$ on the interval $[1.73, 2.32]$, and it was clear that the interval was a little too large to be a correct answer. We used Newton's method to find more accurate endpoints, and came up with the answer $I = (1.74, 2.30)$. Or course, any subinterval of this interval containing $a = 2$ is also a correct answer.

58. We plotted $f(x) = \cos x$ and its linear approximation $L(x) = \frac{1}{2}\pi - x$ on the interval $[0.85, 2.25]$, and it was clear that this interval was a little too large to be a correct answer. We used Newton's method to improve our estimate of the desired endpoints and came up with the answer $I = (0.897, 2.245)$. Of course, any subinterval of this interval containing $a = \pi/2$ is also a correct answer.

Section 4.3

1. $f'(x) = -2x$; f is increasing on $(-\infty, 0)$ and decreasing on $(0, +\infty)$. Matching graph: (c).

4. $f'(x) = \frac{3}{4}x^2 - 3$; $f'(x) = 0$ when $x = \pm 2$; f is increasing on $(-\infty, -2)$ and on $(2, \infty)$, decreasing on $(-2, +2)$. Matching graph: (a).

7. $f(x) = 2x^2 + C$; $5 = f(0) = C$: $f(x) = 2x^2 + 5$.

10. $f(x) = 4\sqrt{x} + C$; $3 = f(0) = C$: $f(x) = 4\sqrt{x} + 3$.

13. $f'(x) = -4x$, so f is increasing on $(-\infty, 0)$ and decreasing on $(0, +\infty)$.

16. $f'(x) = 3x^2 - 12 = 3(x^2 - 4) = 3(x + 2)(x - 2)$. Hence f is increasing for $x > 2$ and for $x < -2$, decreasing for x in the interval $(-2, 2)$.

19. $f'(x) = 12x^3 + 12x^2 - 24x = 12x(x + 2)(x - 1)$, so the only points where $f'(x)$ can change sign are -2, 0, and 1.

22. $f'(x) = 6(x + 2)(x - 1)$, so f is increasing for $x < -2$ and for $x > 1$, decreasing for $-2 < x < 1$.

25. $f(0) = 0$, $f(2) = 0$, f is continuous for $0 \leq x \leq 2$, and $f'(x) = 2x - 2$ exists for $0 < x < 2$. To find the numbers c satisfying the conclusion of Rolle's theorem, we solve $f'(c) = 0$ to find that $c = 1$ is the only such number.

28. Here,

$$f'(x) = \frac{10}{3}x^{-1/3} - \frac{5}{3}x^{2/3} = \frac{10 - 5x}{3x^{1/3}};$$

$f'(x)$ exists for all x in $(0, 5)$ and f is continuous on the interval $0 \leq x \leq 5$ (the only point that might cause trouble is $x = 0$ but the limit of f and its value there are the same). Because $f(0) = 0 = f(5)$, there is a solution c of $f(x) = 0$ in $(0, 5)$, and clearly $c = 2$.

31. Because $f(1) = 2 \neq 0$, this function does not satisfy the hypotheses of Rolle's theorem. Neither does the conclusion hold, for $f'(x) = 4x^3 + 2x = 2x(2x^2 + 1)$ is never zero on $(0, 1)$.

34. First,

$$f'(x) = \frac{1}{2(x - 1)^{1/2}}$$

exists for all $x > 1$, so f satisfies the hypotheses of the mean value theorem for $2 \leq x \leq 5$. To find c, we solve

$$\frac{1}{2(c - 1)^{1/2}} = \frac{(4)^{1/2} - (1)^{1/2}}{5 - 2};$$

thus $2(c - 1)^{1/2} = 3$, and so $c = 13/4$. Note: $2 < c < 5$.

37. First, $f(x) = |x - 2|$ is not differentiable at $x = 2$, so does not satisfy the hypotheses of the mean value theorem on the given interval $1 \leq x \leq 4$. Wherever $f'(x)$ is defined, its value is 1 or -1, but

$$\frac{f(4) - f(1)}{4 - 1} = \frac{2 - 1}{3} = \frac{1}{3}.$$

is never a value of $f'(x)$. So f satisfies neither the hypotheses nor the conclusion of the mean value theorem on the interval $1 \leq x \leq 4$.

40. The function $f(x) = 3x^{2/3}$ is continuous everywhere, but its derivative $f'(x) = 2x^{-1/3}$ does not exist at $x = 0$. Because $f'(x)$ does not exist for *all* x in $(-1, 1)$, this essential hypothesis of the mean value theorem is not satisfied. Moreover, $f'(x)$ is never zero (the average slope of the graph of f on the interval $-1 \leq x \leq 1$), so the conclusion of the theorem also fails to hold.

43. Let $g(x) = x^4 - 3x - 20$. Then $g(2) = -10 < 0$ and $g(3) = 52 > 0$. By the intermediate value property of continuous functions, $g(x)$ has at least one zero in $(2, 3)$. Its derivative $g'(x) = 4x^3 - 3$ is also continuous, and is zero only when $x = (3/4)^{1/3}$, approximately 0.91. So $g'(x)$ does not change sign on the interval $2 \leq x \leq 3$. For $x \geq 2$, $g'(x) = 4x^3 - 3 \geq 4 \cdot 2^3 - 3 = 29 > 0$. Consequently g is increasing on that interval, and so $g(x)$ has at most one zero there. The conclusion is that $g(x) = 0$ has exactly one solution for $2 \leq x \leq 3$.

46. A change of 15 miles per hour in 10 minutes is an average change of 1.5 miles per hour per minute, which is an average change of 90 miles per hour per hour. By the mean value theorem, the instantaneous rate of change of velocity must have been exactly 90 miles per hour per hour at some time in the given 10-minute interval.

49. Because

$$f'(x) = \frac{3}{2}(1 + x)^{1/2} - \frac{3}{2} = \frac{3}{2}\left(\sqrt{1 + x} - 1\right),$$

it is clear that $f'(x) > 0$ for $x > 0$. Also $f(0) = 0$; it follows that $f(x) > 0$ for all $x > 0$. That is,

$$(1 + x)^{3/2} > 1 + \frac{3}{2}x \quad \text{for } x > 0.$$

52. Suppose that $f(x) = 0$ for $x = x_1, x_2, \ldots, x_k$ in the interval $[a, b]$. By Rolle's theorem, $f'(x) = 0$ for some c_1 in (x_1, x_2), some c_2 in (x_2, x_3), $\ldots,$ and some c_{k-1} in (x_{k-1}, x_k). The numbers $c_1, c_2, \ldots, c_{k-1}$ are distinct because they come from disjoint intervals, and this proves the desired result.

55. Let $f(x) = (\tan x)^2$ and let $g(x) = (\sec x)^2$. Then

$$f'(x) = 2(\tan x)(\sec^2 x) \quad \text{and} \quad g'(x) = 2(\sec x)(\sec x \tan x) = f'(x) \text{ on } (-\pi/2, \pi/2).$$

Therefore there exists a constant C such that $f(x) = g(x) + C$ for all x in $(-\pi/2, \pi/2)$. Finally, $f(0) = 0$ and $g(0) = 1$, so $C = f(0) - g(0) = -1$.

58. To show that the graph of $f(x) = x^4 - x^3 + 7x^2 + 3x - 11$ has a horizontal tangent line, we must show that its derivative $f'(x) = 4x^3 - 3x^2 + 14x + 3$ has the value zero at some number c. Now $f'(x)$ is a polynomial, thus is continuous everywhere, and so has the intermediate value property; moreover, $f'(-1) = -18$ and $f'(0) - 3$, so $f'(c) = 0$ for some number c in $(-1, 0)$. (The value of c is approximately -0.203058.)

61. Let $h(x) = 1 - \frac{1}{2}x^2 - \cos x$. Then $h'(x) = -x + \sin x$. By Example 8, $\sin x < x$ for all $x > 0$, so $h'(x) < 0$ for all $x > 0$. If $x > 0$, then $\dfrac{h(x) - 0}{x - 0} = h'(c)$ for some $c > 0$, so $h(x) < 0$ for all $x > 0$; that is, $\cos x > 1 - \frac{1}{2}x^2$ for all $x > 0$.

Section 4.4

1. $f'(x) = 2x - 4$; $x = 2$ is the only critical point. Because $f'(x) > 0$ for $x > 2$ and $f'(x) < 0$ for $x < 2$, it follows that $f(2) = 1$ is the global minimum value of $f(x)$.

4. $f'(x) = 3x^2 - 3 = 3(x+1)(x-1)$, so $x = 1$ and $x = -1$ are the only critical points. If $x < -1$ or if $x > 1$, then $f'(x) > 0$, whereas $f'(x) < 0$ on $(-1, 1)$. So $f(-1) = 7$ is a local maximum value and $f(1) = 3$ is a local minimum.

7. $f'(x) = -6(x-5)(x+2)$; $f'(x) < 0$ if $x < -2$ and if $x > 5$, but $f'(x) > 0$ for $-2 < x < 5$. Hence $f(-2) = -58$ is a local minimum value of $f(x)$ and $f(5) = 285$ is a local maximum value.

10. $f'(x) = 15x^2(x+1)(x-1)$, so $f'(x) > 0$ if $x < -1$ and if $x > 1$, but $f'(x) < 0$ on $(-1, 0)$ and on $(0, 1)$. Therefore $f(0) = 0$ is not an extremum of $f(x)$, but $f(-1) = 2$ is a local maximum value and $f(1) = -2$ is a local minimum value.

13. Here,

$$f'(x) = 2x - \frac{2}{x^2} = \frac{2(x^3 - 1)}{x^2} = \frac{2(x-1)(x^2+x+1)}{x^2}.$$

Because $x^2 + x + 1 > 0$ for all x, the only critical point is $x = 1$; note that f is not defined at $x = 0$. Also $f'(x)$ has the sign of $x - 1$, so $f'(x) > 0$ for $x > 1$ and $f'(x) < 0$ for $0 < x < 1$ and for $x < 0$. Consequently $f(1) = 3$ is a local minimum value of $f(x)$. It is not a global minimum; check the behavior of $f(x)$ for x negative and near zero.

16. Because $f'(x) = \frac{1}{3}x^{-2/3}$, $f'(x) > 0$ for all x except for x except for $x = 0$, where f is continuous but $f'(x)$ is not defined. Consequently f has no extrema. Examination of the behavior of $f(x)$ and $f'(x)$ for x near zero makes it clear that the graph of f has a vertical tangent at $(0, 4)$.

19. $f'(x) = 3\sin^2 x \cos x$; $f'(x) = 0$ when $x = -\pi/2$, 0, or $\pi/2$. f is decreasing on $(-3, -\pi/2)$ and on $(\pi/2, 3)$, increasing on $(-\pi/2, \pi/2)$. So f has a [global] minimum at $(-\pi/2, -1)$ and a [global] maximum at $(\pi/2, 1)$.

22. $f'(x) = 3\tan^2 x \sec^2 x > 0$ on $(-1, 1)$; no extrema.

25. $f'(x) = 2\sec^2 x - 2\tan x \sec^2 x = (2\sec^2 x)(1 - \tan x)$. $f'(x) = 0$ when $x = \pi/4$; f is increasing on $(0, \pi/4)$ and decreasing on $(\pi/4, 1)$; maximum at $(\pi/4, 1)$.

28. We assume that the length turned upward is the same on each side—call it y. If the width of the gutter is x, then we have the constraint $xy = 18$, and we are to minimize the width $x + 2y$ of the strip. Its width is given by the function

$$f(x) = x + \frac{36}{x}, \qquad x > 0,$$

for which

$$f'(x) = 1 - \frac{36}{x^2}.$$

The only critical point in the domain of f is $x = 6$, and if $0 < x < 6$ then $f'(x) < 0$, whereas $f'(x) > 0$ for $x > 6$. Thus $x = 6$ yields the minimum value $f(6) = 12$ of the function f. Answer: The minimum possible width of the strip is 12 inches.

31. Base of box: x wide, $2x$ long. Height: y. Then the box has volume $2x^2 y = 972$, so $y = 486x^{-2}$. Its total surface area is $A = 2x^2 + 6xy$, so we minimize

$$A = A(x) = 2x^2 + \frac{2916}{x}, \qquad x > 0.$$

Now

$$A'(x) = 4x - \frac{2916}{x^2},$$

so the only critical point of $A(x)$ occurs when $4x^3 = 2916$; that is, when $x = 9$. It is easy to verify that $A'(x) < 0$ for $0 < x < 9$ and that $A'(x) > 0$ for $x > 9$. Therefore $A(9)$ is the global minimum value of $A(x)$. Answer: The dimensions of the box are 9 inches wide, 18 inches long, 6 inches high.

34. If $(x, y) = (x, 4 - x^2)$ is a point on the parabola $y = 4 - x^2$, then the square of its distance from the point $(3, 4)$ is
$$h(x) = (x - 3)^2 + (4 - x^2 - 4)^2 = (x - 3)^2 + x^4.$$

We minimize the distance by minimizing its square:
$$h'(x) = 2(x - 3) + 4x^3;$$

$h'(x) = 0$ when $2x^3 + x - 3 = 0$. It is clear that $h'(1) = 0$, so $x - 1$ is a factor of $h'(x)$; $h'(x) = 0$ is equivalent to $(x - 1)(2x^2 + 2x + 3) = 0$. The quadratic factor in the last equation is always positive, so $x = 1$ is the only critical point of $h(x)$. Also $h'(x) < 0$ if $x < 1$, whereas $h'(x) > 0$ for $x > 1$, so $x = 1$ yields the global minimum value $h(1) = 5$ for $h(x)$. When $x = 1$ we have $y = 3$, so the point on the parabola $y = 4 - x^2$ closest to $(3, 4)$ is $(1, 3)$, at distance $\sqrt{5}$ from it.

37. Let the square base of the box have edge length x and let its height be y, so that its total volume is $x^2 y = 62.5$ and the surface area of this box-without-top will be $A = x^2 + 4xy$. So
$$A = A(x) = x^2 + \frac{250}{x}, \qquad x > 0.$$

Now
$$A'(x) = 2x - \frac{250}{x^2},$$

so $A'(x) = 0$ when $x^3 = 125$: $x = 5$. In this case, $y = 2.5$. Also $A'(x) < 0$ if $0 < x < 5$ and $A'(x) > 0$ if $x > 5$, so we have found the global minimum for $A(x)$. Answer: Square base of edge length 5 inches, height 2.5 inches.

40. If the print width is x and its height is y (in inches), then the page area is $A = (x + 2)(y + 4)$. We are to minimize A given $xy = 30$. Because $y = 30/x$,
$$A = A(x) = 4x + 38 + \frac{60}{x}, \qquad x > 0.$$

Now
$$A'(x) = 4 - \frac{60}{x^2};$$

$A'(x) = 0$ when $x = \sqrt{15}$. But $A'(x) > 0$ for $x > \sqrt{15}$ whereas $A'(x) < 0$ for $x < \sqrt{15}$. Therefore $x = \sqrt{15}$ yields the global minimum value of $A(x)$, which is $38 + 8\sqrt{15}$, approximately 68.98 square inches.

43. If the dimensions of the rectangle are x by y, and the line segment bisects the side of length x, then the square of the length of the segment is
$$f(x) = \left(\frac{x}{2}\right)^2 + y^2 = \frac{x^2}{4} + \frac{4096}{x^2}, \qquad x > 0,$$

because $y = 64/x$. Now
$$f'(x) = \frac{x}{2} - \frac{8192}{x^3}.$$

When $f'(x) = 0$, we must have $x = +8\sqrt{2}$, so that $y = 4\sqrt{2}$. We have found the minimum of f because if $0 < x < 8\sqrt{2}$ then $f'(x) < 0$, and $f'(x) > 0$ if $x > 8\sqrt{2}$. The minimum length satisfies $L^2 = f\left(8\sqrt{2}\right)$, so that $L = 8$ centimeters.

46. By similar triangles, $y/1 = 8/x$, and
$$L_1 + L_2 = L = \left[(x + 1)^2 + (y + 8)^2\right]^{1/2}.$$

We minimize L by minimizing

$$f(x) = L^2 = (x+1)^2 + \left(8 + \frac{8}{x}\right)^2, \qquad x > 0.$$

$$f'(x) = 2 + 2x - \frac{128}{x^3} - \frac{128}{x^2};$$

$f'(x) = 0$ when $2x^3 + 2x^4 - 128 - 128x = 0$, which leads to the equation

$$(x+1)(x-4)(x^2 + 4x + 16) = 0.$$

The only relevant solution is $x = 4$. Because $f'(x) < 0$ for x in the interval $(-1, 4)$ and $f'(x) > 0$ if $x > 4$, we have indeed found the global minimum of f. The corresponding value of y is 2, and the length of the shortest ladder is $L = 5\sqrt{5}$ feet, approximately 11 ft 2 in.

49. Let z be the length of the segment from the top of the tent to the midpoint of one side of its base. Then $x^2 + y^2 = z^2$. The total surface area of the tent is

$$A = 4x^2 + (4)(\tfrac{1}{2})(2x)(z) = 4x^2 + 4xz = 4x^2 + 4x(x^2 + y^2)^{1/2}.$$

Because the (fixed) volume V of the tent is given by

$$V = \frac{1}{3}(4x^2)(y) = \frac{4}{3}x^2 y,$$

we have $y = 3V/(4x^2)$, so

$$A = A(x) = 4x^2 + \frac{1}{x}(16x^6 + 9V^2)^{1/2}.$$

After simplifications, the condition $dV/dx = 0$ takes the form

$$8x\left(16x^6 + 9V^2\right)^{1/2} - \frac{1}{x^2}\left(16x^6 + 9V^2\right) + 48x^4 = 0,$$

which has solution $x = 2^{-7/6}\sqrt[3]{3V}$. Because this is the only positive solution of the equation, and because it is clear that neither large values of x nor values of x near zero will yield small values of the surface area, this is the desired value of x.

Section 4.5

1. $f(x) \to +\infty$ as $x \to +\infty$, $f(x) \to -\infty$ as $x \to -\infty$. Matching graph: 4.5.13(c).

4. $f(x) \to -\infty$ as $x \to +\infty$, $f(x) \to -\infty$ as $x \to -\infty$. Matching graph: 4.5.13(b).

10. The critical points occur where $x = -\frac{8}{3}$, where $x = 0$, and where $x = \frac{5}{2}$. The graph is increasing on $(-\infty, -\frac{8}{3})$ and on $(0, \frac{5}{2})$. The graph is decreasing on $(-\frac{8}{3}, 0)$ and on $(\frac{5}{2}, +\infty)$.

16. $f'(x) = -8 - 4x$ is positive for $x < -2$, negative for $x > -2$. The graph is a parabola opening downward, vertical axis, vertex (and global maximum) at $(-2, 13)$.

22. $f'(x) = 3(x+3)(x-3)$: Local maximum at $(-3, 54)$, local minimum at $(3, -54)$.

28. $f'(x) = 4x^2(x+3)$ is positive for $x > -3$ and negative for $x < -3$; there is a horizontal tangent but no extremum at $x = 0$. There is a global minimum at $(-3, -27)$.

31. $f'(x) = 4x - 3$ changes sign at $x = 3/4$. The function is increasing for $x > 3/4$ and decreasing for $x < 3/4$; it has a global minimum at $(3/4, -81/8)$.

34. $f'(x) = 3x^2 + 4$ is positive for all x, so the function is increasing for all x; there are no extrema, and $(0, 0)$ is the only intercept.

40. $f'(x) = 15(x+1)(x-1)(x+2)(x-2)$, so f is increasing if $x < -2$, if $-1 < x < 1$, and if $x > 1$, decreasing if $-2 < x < 1$ and if $1 < x < 2$. So there are local maxima at $(-2, -16)$ and $(1, 38)$, local minima at $(-1, -38)$ and $(2, 16)$. The only intercept is $(0, 0)$.

46. In this case
$$f'(x) = \frac{8(x+2)(x-2)}{3x^{1/3}},$$
which is positive for $x > 2$ and for $-2 < x < 0$, negative for $x < -2$ and for $0 < x < 2$. Note that f is continuous at $x = 0$ even though $f'(0)$ does not exist. Moreover, for x near zero, we have
$$f(x) \approx -16x^{2/3} \quad \text{and} \quad f'(x) \approx -\frac{32}{3x^{1/3}}.$$
Consequently $f'(x) \to -\infty$ as $x \to 0^+$, while $f'(x) \to +\infty$ as $x \to 0^-$. This is consistent with the observation that $f(x) < 0$ for all x near (but not equal to) zero. The origin is a local maximum and there are global minima where $|x| = 2$.

49. The graph is shown below.

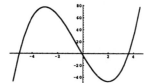

52. The graph is shown below.

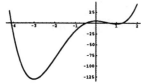

55. $x^3 - 3x + 3 \approx (x + 2.1038)(x^2 - 2.1038x + 1.4260)$.

Section 4.6

1. $f'(x) = 8x^3 - 9x^2 + 6$, $f''(x) = 24x^2 - 18x$, $f'''(x) = 48x - 18$

4. $g'(t) = 2t + \frac{1}{2}(t+1)^{-1/2}$, $g''(t) = 2 - \frac{1}{4}(t+1)^{-3/2}$, $g'''(t) = \frac{3}{8}(t+1)^{-5/2}$

10. $h'(z) = \dfrac{8z}{(z^2+4)^2}$, $h''(z) = \dfrac{32 - 24z^2}{(z^2+4)^3}$, $h'''(z) = \dfrac{96z^3 - 384z}{(z^2+4)^4}$

16. Given: $x^2 + y^2 = 4$.
$$2x + 2yy'(x) = 0, \quad \text{so} \quad y'(x) = -\frac{x}{y}.$$
$$y''(x) = -\frac{y - xy'(x)}{y^2} = -\frac{y + (x^2/y)}{y^2} = -\frac{y^2 + x^2}{y^3} = -\frac{4}{y^3}.$$

22. $\sin^2 x + \cos^2 y = 1$: $2\sin x \cos x = 2y'(x)\sin y \cos y = 0$; $y'(x) = \dfrac{\sin x \cos x}{\sin y \cos y}$.

$\dfrac{d^2 y}{dx^2}$ can be simplified (with the aid of the original equation) to
$$\frac{\cos^2 x \sin^2 y - \sin^2 x \cos^2 y}{\sin^3 y \cos^3 y} \equiv 0 \quad \text{if } y \text{ is not an integral multiple of } \pi/2.$$

28. Critical point: $(7.5, -1304.69)$; inflection points: $(0, -250)$ and $(5, -875)$

34. $f'(x) = 3x(x - 2)$; $f''(x) = 6(x - 1)$. There are critical points at $(0, 0)$ and at $(2, -4)$. Now $f''0) = -6 < 0$, so there is a local maximum at $(0, 0)$; $f''(2) = 6 > 0$, so there is a local minimum at $(2, -4)$. The only possible inflection point is $(1, -2)$, and it is indeed an inflection point because f'' changes sign there.

40. $f'(x) = x^2(x + 2)(5x + 6)$ and $f''(x) = 4x(5x^2 + 12x + 6)$. So the critical points occur where $x = 0$, $x = -2$, and $x = -6/5$. Now $f''(0) = 0$, so the second derivative test fails here, but $f'(x) > 0$ for x near zero but $x \neq 0$, so $(0, 0)$ is not an extremum for the graph of f. Next, $f''(-2) = -16 < 0$, so $(-2, 0)$ is a local maximum point; $f''(-6/5) = 5.76 > 0$, so $(-6/5, -3456/3125)$ is a local minimum point. The possible inflection points occur at

$$x = 0, \quad x = \tfrac{1}{5}\left(-6 + \sqrt{6}\right), \quad \text{and} \quad \tfrac{1}{5}\left(-6 - \sqrt{6}\right).$$

In decimal form these are $x = 0$, $x \approx -0.710$, and $x \approx -1.690$. Because $f''(-2) = -16 < 0$, $f''(-1) = 4 > 0$, $f''(-0.5) = -2.5 < 0$, and $f''(1) = 92 > 0$, each of the three numbers displayed above is the abscissa of an inflection point of the graph of f.

46. $f(x) = \sin^3 x$, $-\pi < x < \pi$:

$$f'(x) = 3\sin^2 x \cos x,$$
$$f''(x) = 6\sin x \cos^2 x - 3\sin^3 x = 3(2\cos^2 x - \sin^2 x)\sin x.$$

Minimum at $(-\pi/2, -1)$; maximum at $(\pi/2, 1)$; inflection points at $x = 0$ and at the four solutions of $\tan^2 x = 2$ in $(-\pi, \pi)$.

52. We assume that the length turned upward is the same on each side—call it y. If the width of the gutter is x, then we have the constraint $xy = 18$, and we are to minimize the width $x + 2y$ of the strip. Its width is given by the function

$$f(x) = x + \frac{36}{x}, \qquad x > 0,$$

for which

$$f'(x) = 1 - \frac{36}{x^2} \quad \text{and} \quad f''(x) = \frac{72}{x^3}.$$

The only critical point in the domain of f is $x = 6$, and $f''(x) > 0$ on the entire domain of f. Consequently the graph of f is concave upward for all $x > 0$. Because f is continuous for such x, $f(6) = 12$ is the global minimum of f.

55. Base of box: x wide, $2x$ long. Height: y. Then the box has volume $2x^2 y = 972$, so $y = 486x^{-2}$. Its total surface area is $A = 2x^2 + 6xy$, so we minimize

$$A = A(x) = 2x^2 + \frac{2916}{x}, \qquad x > 0.$$

Now

$$A'(x) = 4x - \frac{2916}{x^2} \quad \text{and} \quad A''(x) = 4 + \frac{5832}{x^3}.$$

The only critical point of A occurs when $x = 9$, and $A''(x)$ is always positive. So the graph of $y = A(x)$ is concave upward for all $x > 0$; consequently, $(9, A(9))$ is the global minimum point of A. Answer: The dimensions of the box are 9 inches wide, 18 inches long, and 6 inches high.

58. Let x denote the length of each side of the square base of the solid and let y denote its height. Then its total volume is $x^2 y = 1000$. We are to minimize its total surface area $A = 2x^2 + 4xy$. Now $y = 1000/(x^2)$, so

$$A = A(x) = 2x^2 + \frac{4000}{x}, \qquad x > 0.$$

Therefore

$$A'(x) = 4x - \frac{4000}{x^2} \quad \text{and} \quad A''(x) = 4 + \frac{8000}{x^3}.$$

The only critical point occurs when $x = 10$, and $A''(x) > 0$ for all x in the domain of A, so $x = 10$ yields the global minimum value of $A(x)$. In this case, $y = 10$ as well, so the solid is indeed a cube.

61. Let x denote the radius and y the height of the cylinder (in inches). Then its cost (in cents) is $C = 8\pi x^2 + 4\pi xy$, and we also have the constraint $\pi x^2 y = 100$. So

$$C = C(x) = 8\pi x^2 + \frac{400}{x}, \qquad x > 0.$$

Now

$$C'(x) = 16\pi x - \frac{400}{x^2} \quad \text{and} \quad C''(x) = 16\pi + \frac{800}{x^3}.$$

The only critical point in the domain of C is $x = \sqrt[3]{25/\pi}$ (about 1.9965 inches) and, consequently, when $y = \sqrt[3]{1600/\pi}$ (about 7.9859 inches). Because $C''(x) > 0$ for all x in the domain of C, we have indeed found the dimensions that minimize the cost of the can. For simplicity, note that $y = 4x$ at the minimum: The height of the can is twice its diameter.

64. Given: $f(x) = 3x^4 - 4x^3 - 5$. Then

$$f'(x) = 12x^3 - 12x^2 = 12x^2(x-1) \quad \text{and} \quad f''(x) = 36x^2 - 24x = 12x(3x-2).$$

So the graph of f is increasing for $x > 1$ and decreasing for $x < 1$ (even though there's a horizontal tangent at $x = 0$), concave upward for $x < 0$ and $x > 2/3$, concave downward on $(0, 2/3)$. There is a global minimum at $(1, -6)$, inflection points at $(0, -5)$ and at $(2/3, -151/27)$. The x-intercepts are approximately -0.906212 and 1.682971.

70. Given: $f(x) = (x-1)^2(x+2)^3$. Then

$$f'(x) = (x-1)(x+2)^2(5x+1) \quad \text{and} \quad f''(x) = 2(x+2)(10x^2 + 4x - 5).$$

The zeros of $f''(x)$ are $x = -2$, $x \approx 0.535$, and $x \approx -0.935$. It follows that $(1, 0)$ is a local minimum (from the second derivative test), that $(-0.2, 8.39808)$ is a local maximum, and that $(-2, 0)$ is not an extremum. Also, the second derivative changes sign at each of its zeros, so each of these three zeros is the abscissa of an inflection point on the graph.

76. Given: $f(x) = x^{1/3}(6-x)^{2/3}$. Then

$$f'(x) = \frac{2-x}{x^{2/3}(6-x)^{1/3}} \quad \text{and} \quad f''(x) = -\frac{8}{x^{5/3}(6-x)^{4/3}}.$$

If $|x|$ is large, then $(6-x)^{2/3} \approx x^{2/3}$, so $f(x) \approx x$ for such x. This aids in sketching the graph, which has a local maximum where $x = 2$, a local minimum at $(6, 0)$, vertical tangents at $(6, 0)$ and at the origin. It is increasing for $x < 2$ and for $x > 6$, decreasing on the interval $(2, 6)$, concave upward for $x < 0$, and concave downward on $(0, 6)$ and for $x > 6$. All the intercepts have been mentioned, too.

82. Figure 4.6.33(a)

85. $\dfrac{dz}{dx} = \dfrac{dz}{dy} \cdot \dfrac{dy}{dx}$. So $\dfrac{d^2z}{dx^2} = \dfrac{dz}{dy} \cdot \dfrac{d^2y}{dx^2} + \dfrac{dy}{dx} \cdot \dfrac{d^2z}{dy^2} \cdot \dfrac{dy}{dx}$.

88. If $f(x) = Ax^4 + Bx^3 + Cx^2 + Dx + E$, then both $f'(x)$ and $f''(x)$ are continuous for all x, and $f''(x) = 12Ax^2 + 6Bx + 2C$. In order for $f''(x)$ to change sign, we must have $f''(x) = 0$. If so, then (because $f''(x)$ is a quadratic polynomial) either the graph of $f''(x)$ crosses the x-axis in two places or is tangent to it at a single point. In the first case, $f''(x)$ changes sign twice, so there are two points of inflection on the graph of f. In the second case, $f''(x)$ does not change sign, so f has no inflection points. Therefore the graph of a polynomial of degree four has either exactly two inflection points or else none at all.

Section 4.7

1. $\dfrac{x}{x+1} = \dfrac{1}{1+(1/x)} \to 1$ as $x \to +\infty$.

4. The numerator approaches -2 as $x \to -1$, whereas the denominator approaches zero. Therefore this limit does not exist.

7. The numerator is equal to the denominator for all $x \neq -1$, so the limit is 1.

10. Divide each term in numerator and denominator by $x^{3/2}$, the highest power of x that appears in any term. The numerator then becomes $2x^{-1/2} + x^{-3/2}$, which approaches 0 as $x \to +\infty$; the denominator becomes $x^{-1/2} - 1$, which approaches -1 as $x \to +\infty$. Therefore the limit is 0.

13. $\dfrac{4x^2 - x}{x^2 + 9} = \dfrac{4 - (1/x)}{1 + (9/x^2)} \to 4$ as $x \to +\infty$, so the limit is $\sqrt{4} = 2$.

16.
$$\lim_{x \to -\infty} \left(2x - \sqrt{4x^2 - 5x}\right) = \lim_{x \to -\infty} \frac{4x^2 - (4x^2 - 5x)}{2x + \sqrt{4x^2 - 5x}} = \lim_{x \to -\infty} \frac{5x}{2x + \sqrt{4x^2 - 5x}}$$
$$= \lim_{x \to -\infty} \frac{5}{2 + \left(\dfrac{\sqrt{4x^2 - 5x}}{-\sqrt{x^2}}\right)} = \lim_{x \to -\infty} \frac{5}{2 - \sqrt{4 - (5/x)}} = -\infty.$$

22. Matches 4.7.9 (c)

28. Matches 4.7.9 (e)

31. Here we have
$$f'(x) = -\frac{6}{(x+2)^3} \quad \text{and} \quad f''(x) = \frac{18}{(x+2)^4}.$$

The graph is increasing and concave upward for $x < -2$, decreasing and concave upward for $x > -2$. There are no extrema and no inflection points; the y-intercept is $(0, \frac{3}{4})$. The line $x = -2$ is a vertical asymptote and the x-axis is a horizontal asymptote.

34. Here,
$$f'(x) = -\frac{2}{(x-1)^2} \quad \text{and} \quad f''(x) = \frac{4}{(x-1)^3}.$$

The graph is decreasing and concave downward if $x < 1$, decreasing and concave upward if $x > 1$. There are no extrema and no inflection points; the y-intercept is $(0, -1)$ and the only x-intercept is $(-1, 0)$. The line $x = 1$ is a vertical asymptote and the line $y = 1$ is a horizontal asymptote.

37.
$$f'(x) = -\frac{2x}{(x^2 - 9)^2} \quad \text{and} \quad f''(x) = \frac{6(x^2 + 3)}{(x^2 - 9)^3}.$$

The graph is increasing for $x < -3$ and for $-3 < x < 0$; the graph is decreasing for $0 < x < 3$ and for $3 < x$. It is concave upward if $x < -3$ and if $x > 3$, concave downward on $(-3, 3)$. The only extremum is the local maximum and y-intercept at the point $(0, -\frac{1}{9})$, and there are no points of inflection. The x-axis is a horizontal asymptote and the lines $x = -3$ and $x = 3$ are vertical asymptotes.

40.
$$f'(x) = -\frac{2(2x - 1)(x + 1)}{x^2(x - 2)^2} \quad \text{and} \quad f''(x) = \frac{2\left(4x^3 + 3x^2 - 6x + 4\right)}{(x^2 - 2x)^3}.$$

The zeros of $f'(x)$ occur at -1 and 0.5, and the only zero of $f''(x)$ is at $x \approx -1.8517$. The graph is decreasing for $x < -1$, for $0.5 < x < 2$, and for $x > 2$; increasing for $-1 < x < 0$ and for $0 < x < 0.5$. It is concave downward for $x < -1.85^*$ and on $(0, 2)$, concave upward for $x > 2$ and on $(-1.85^*, 0)$. The line $y = 2$ is a horizontal asymptote and the lines $x = 0$ and $x = 2$ are vertical asymptotes.

There is a local minimum at $(-1, 1)$, a local maximum at $(0.5, -2)$, and a point of inflection at $(-1.85^*, 1.10^*)$. (*Coordinates approximate.)

43.

$$f'(x) = \frac{x(x-2)}{(x-1)^2} \quad \text{and} \quad f''(x) = \frac{2}{(x-1)^3}.$$

The graph is increasing for $x < 0$ and for $x > 2$, decreasing on $(0, 1)$ and on $(1, 2)$. It is concave upward if $x > 1$, concave downward if $x < 1$. The origin is a local maximum and the only intercept, the only other extremum is a local minimum at $(2, 4)$, and there are no inflection points. The lines $y = x + 1$ and $x = 1$ are asymptotes.

46.

$$f'(x) = -\frac{2x}{(x^2 - 4)^2} \quad \text{and} \quad f''(x) = \frac{6x^2 + 8}{(x^2 - 4)^3}.$$

The graph is increasing for $x < -2$ and on $(-2, 0)$, decreasing on $(0, 2)$ and for $x > 2$. It is concave upward if $|x| > 2$ and concave downward if $|x| < 2$. The only intercept is at the point $(0, -0.25)$, which is also a local maximum; there are no other extrema and no inflection points. The two vertical lines $x = -2$ and $x = 2$ are asymptotes, as is the x-axis.

49.

$$f(x) = \frac{1}{x^2 - x - 2} = \frac{1}{(x-1)(x-2)} : \quad f'(x) = -\frac{2x - 1}{(x+1)^2(x-2)^2} \quad \text{and} \quad f''(x) = \frac{6(x^2 - x + 1)}{(x+1)^3(x-2)^3}.$$

It is useful to notice the discontinuities at $x = -1$ and at $x = 2$, that $f(x) \to 0$ as $x \to +\infty$ and as $x \to -\infty$, and that $f(x) > 0$ for $x > 2$ and for $x < -1$ while $f(x) < 0$ on the interval $(-1, 2)$. The denominator in $f'(x)$ is never negative, so the graph is decreasing for $\frac{1}{2} < x < 2$ and for $2 < x$ and increasing for $x < -1$ and for $-1 < x < \frac{1}{2}$. The graph is concave upward for $x > 2$ and for $x < -1$, concave downward on $(-1, 2)$. There are no inflection points or intercepts, but there is a local maximum at $(\frac{1}{2}, -\frac{4}{9})$. The x-axis is a horizontal asymptote and the lines $x = -1$ and $x = 2$ are vertical asymptotes.

52.

$$f'(x) = -\frac{x^2 + 1}{(x^2 - 1)^2} \quad \text{and} \quad f''(x) = \frac{2x(x^2 + 3)}{(x^2 - 1)^3}.$$

The graph is decreasing on $(-\infty, -1)$, on $(-1, 1)$, and on $(1, \infty)$; it is concave upward for $x > 1$ and for $-1 < x < 0$, concave downward for $0 < x < 1$ and for $x < -1$. The only intercept is $(0, 0)$, which is also an inflection point. Note that $f(x) \to 0$ as $|x| \to \infty$, so the x-axis is a horizontal asymptote; $f(x) \to +\infty$ as $x \to -1^+$ and as $x \to +1^+$, while $f(x) \to -\infty$ as $x \to -1^-$ and as $x \to +1^-$, so the lines $x = -1$ and $x = 1$ are vertical asymptotes. There are no extrema.

55. The x-axis is a horizontal asymptote, and there are vertical asymptotes at $x = 0$ and $x = 2$. There are local minima at $(-1.9095, -0.3132)$ and $(1.3907, 3.2649)$ and a local maximum at $(4.5188, 0.1630)$ (all coordinates approximate, of course), and inflection points at $(-2.8119, -0.2768)$ and $(6.0623, 0.1449)$.

58. The horizontal line $y = 1$ is an asymptote, as are the vertical lines $x = 0$ and $x = 2$. There are local maxima at $(-5.6056, 1.1726)$ and $(1.6056, -8.0861)$, $(-1.2994, 0.2289)$, and $(3.6777, 0.1204)$. (Numbers with decimal points are approximations.)

61. The x-axis is a horizontal asymptote; there are vertical asymptotes at $x = -0.5321$, $x = 0.6527$, and $x = 2.8794$. There is a local minimum at $(0, 0)$ and a local maximum at $(\sqrt[3]{2}, -0.9008)$. There are no inflection points (Numbers with decimal points are approximations.)

64. The line $2y = x$ is a slant asymptote in both the positive and negative directions; thus there is no horizontal asymptote. There is a vertical asymptote at $x = -1.4757$. There is a local maximum at $(-2.9821, -2.1859)$ and a local minimum at $(0.7868, -2.8741)$. There are inflection points at

$(-0.2971, 0.7736)$, $(0.5713, 0.5566)$, $(1, -2)$, and $(9.1960, 4.6515)$. (Numbers with decimal points are approximations.)

67. The line $2y = x$ is a slant asymptote in both the positive and negative directions; thus there is no horizontal asymptote. There is a vertical asymptote at $x = -1.7277$. There are local maxima at $(-3.1594, -2.3665)$ and $(1.3381, 1.7792)$, local minima at $(-0.5379, -0.3591)$ and $(1.8786, 1.4388)$. There are inflection points at $(0, 0)$, $(0.5324, 0.4805)$, $(1.1607, 1.4294)$, and $(1.4627, 1.6727)$. (Numbers with decimal points are approximations.)

70. Because $f(x) \approx x^3$ when $|x|$ is large, we obtain the graph of f by making "modifications" in the graph of $y = x^3$ at and near the discontinuity of $f(x)$ at $x = 1$. We are aided in sketching the graph of f by finding its x-intercepts—these are approximately -1.654 and 2.172—as well as its y-intercept 12 and its inflection points $(2.22, 1.04)$ and $(-0.75, 6.44)$ (also approximations).

Chapter 4 Miscellaneous

1. $dy = \frac{3}{2}(4x - x^2)^{1/2}(4 - 2x)\, dx$

4. $dy = 2x \cos(x^2)\, dx$

7. Let $f(x) = x^{1/2}$; $f'(x) = \frac{1}{2}x^{-1/2}$. Then

$$\sqrt{6401} = f(6400 + 1) \approx f(6400) + 1 \cdot f'(6400)$$
$$= 80 + \frac{1}{160} = \frac{12801}{160} = 80.00625.$$

(Actually, $\sqrt{6401} \approx 80.00624976$.)

10. $\sqrt[3]{999} = \sqrt[3]{1000} - 1 \cdot \frac{1}{3}(1000)^{-2/3} = 10 - \frac{1}{300} = \frac{2999}{300} \approx 9.996667$.

16. With $f(x) = x^{1/10}$, $f'(x) = \frac{1}{10}x^{-9/10}$, $x = 1024$, and $\Delta x = -24$, we obtain

$$(1000)^{1/10} = f(x + \Delta x) \approx f(x) + f'(x)\, \Delta x$$
$$= (1024)^{1/10} + (-24)\left(\frac{1}{10}\right)(1024)^{-9/10}$$
$$= 2 - \frac{3}{640} = \frac{1277}{640} \approx 1.9953.$$

22. Here, $dL = (-13)(10^{30})E^{-14}\, dE$. We take $E = 110$ and $\Delta E = +1$ and obtain

$$dL \approx (-13)(10^{30})\left(110^{-14}\right)(+1) \approx -342 \text{ hours.}$$

The actual decrease is $L(110) - L(111) \approx 2896.6 - 2575.1 \approx 321.5$ (hours).

28. $\dfrac{f(4) - f(0)}{4 - 0} = f'(c)$: $\dfrac{2 - 0}{4} = \frac{1}{2}c^{-1/2}$; $c^{-1/2} = 1$; $c = 1$.

31. $f'(x) = 15(x^4 - x^2 + 4) > 0$ for all x; $f''(x) = 30x(2x^2 - 1)$, so the graph is concave upward for $-1/\sqrt{2} < x < 0$ and for $x > 1/\sqrt{2}$, concave downward for $x < -1/\sqrt{2}$ and for $0 < x < 1/\sqrt{2}$. There are no extrema, there are inflection points where $f''(x) = 0$, and $(0, 0)$ is the only intercept.

34. Let $g(x) = x^5 + x - 5$. Then $g(2) = 29 > 0$ while $g(1) = -3 < 0$. Because $g(x)$ is a polynomial, it has the intermediate value property. Therefore the equation $g(x) = 0$ has at least one solution in the interval $1 \leq x \leq 2$. Moreover, $g'(x) = 5x^4 + 1$, so $g'(x) > 0$ for all x. Consequently g is increasing on the set of all real numbers, and so takes on each value—including zero—at most once. We may conclude that the equation $g(x) = 0$ has exactly one solution, and hence that the equation $x^5 + x = 5$ has exactly one solution. (The solution is approximately 1.299152792.)

40. $g'(x) = -\dfrac{2x}{(x^2+9)^2}$, $g''(x) = \dfrac{6x^2-18}{(x^2+9)^3}$, and $g'''(x) = \dfrac{216x-24x^3}{(x^2+9)^4}$.

46. $\dfrac{dy}{dx} = \dfrac{3y-4x}{10y-3x}$; $\dfrac{d^2y}{dx^2} = -\dfrac{1550}{(10y-3x)^3}$.

52. $(x^2-y^2)^2 = 4xy$:

$$2(x^2-y^2)(2x - 2y\frac{dy}{dx}) = 4x\frac{dy}{dx} + 4y$$

$$(x^2-y^2)\cdot x - (x^2-y^2)\cdot y\frac{dy}{dx} = x\frac{dy}{dx} + y$$

$$(x + x^2y - y^3)\frac{dy}{dx} = x^3 - xy^2 - y$$

$$\frac{dy}{dx} = \frac{x(x^2-y^2)-y}{x + y(x^2-y^2)} = \frac{x^3 - xy^2 - y}{x + x^2y - y^3}.$$

$$\frac{d^2y}{dx^2} = \frac{(x+x^2y-y^3)\left(3x^2 - x2y\frac{dy}{dx} - y^2 - \frac{dy}{dx}\right) - (x^3 - xy^2 - y)\left(1 + x^2\frac{dy}{dx} + 2xy - 3y^2\frac{dy}{dx}\right)}{(x + x^2y - y^3)^2},$$

which upon simplification and substitution for dy/dx becomes:

$$\frac{d^2y}{dx^2} = \frac{3xy(2-xy)}{(x + x^2y - y^3)^3}.$$

55. $f'(x) = 2x^3(3x^2-4)$ and $f''(x) = 6x^2(5x^2-4)$.

58. $f'(x) = \dfrac{3}{(x+2)^2}$ and $f''(x) = -\dfrac{6}{(x+2)^3}$ There are no critical points and no inflection points. The graph is increasing except at the discontinuity at $x = -2$. It is concave upward for $x < -2$ and concave downward for $x > -2$. The vertical line $x = -2$ and the horizontal line $y = 1$ are asymptotes.

61. $f(x) = \dfrac{2x^2}{(x-2)(x+1)}$, $f'(x) = -\dfrac{2x(x+4)}{(x-2)^2(x+1)^2}$, and $f''(x) = \dfrac{4(x^3+6x^2+4)}{(x^2-x-2)^3}$.

64. Here we have $f(x) = x^2(x^2-2)$, $f'(x) = 4x(x+1)(x-1)$, and $f''(x) = 4(3x^2-1)$. So there are intercepts at $(-\sqrt{2},0)$, $(0,0)$, and $(\sqrt{2},0)$. The graph is increasing on the intervals $(1,\infty)$ and $(-1,0)$, decreasing on the intervals $(-\infty,-1)$ and $(0,1)$. It is concave upward where $x^2 > 1/3$ and concave downward where $x^2 < 1/3$. There are global minima at $(-1,-1)$ and $(1,-1)$ and a local maximum at the origin. There are inflection points at the two points where $x^2 = 1/3$.

67. $f'(x) = 3x(4-x)$ and $f''(x) = 6(2-x)$. The abscissas of the x-intercepts are approximately -1.18014, 1.48887, and 5.69127.

70. We have $f(x) = x^2(x^2-12)$, $f'(x) = 4x(x^2-6)$, and $f''(x) = 12(x^2-2)$. The analysis, results, and graph are qualitatively the same as in the solution of Problem 64.

73. The given function $f(x)$ is expressed as a fraction with constant numerator, so we maximize $f(x)$ by minimizing its denominator $(x+1)^2 + 1$. It is clear that $x = -1$ does the trick, so the maximum value of $f(x)$ is $f(-1) = 1$.

76. Let x represent the edge length of the square base of the box. Because the volume of the box is 324, the box has height $324/x^2$. We minimize the cost C of materials to make the box, where

$$C = C(x) = 3x^2 + 4\cdot x \cdot \frac{324}{x^2} = 3x^2 + \frac{1296}{x}, \qquad 0 < x < \infty.$$

Now $C'(x) = 6x - (1296/x^2)$, so $C(x) = 0$ when $x = \sqrt[3]{1296/6} = 6$. Because $C''(6) > 0$, the cost C is minimized when $x = 6$. The box we seek has a square base 6 in. on a side and height 9 in.

79. If the speed of the truck is v, then the trip time is $T = 1000/v$. So the resulting cost is

$$C(v) = \frac{10000}{v} + (1000)\left(1 + (0.0003)v^{3/2}\right),$$

so that

$$\frac{C(v)}{1000} = \frac{10}{v} + 1 + (0.0003)v^{3/2}.$$

Thus

$$\frac{C'(v)}{1000} = -\frac{10}{v^2} + \frac{3}{2}(0.0003)\sqrt{v}.$$

Then $C'(v) = 0$ when $v = (200,000/9)^{2/5} \approx 54.79$ mi/h. This clearly minimizes the cost, since $C''(v) > 0$ for *all* $v > 0$.

82. Let x represent the length of the internal divider. Then the field is x by $2400/x$ ft. We minimize the total length of fencing, given by:

$$f(x) = 3x + \frac{4800}{x}, \qquad 0 < x < \infty.$$

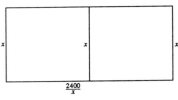

Now $f'(x) = 3 - \frac{4800}{x^2}$, which is zero only when $x = \sqrt{1600} = 40$.
Verification: $f'(x) > 0$ if $x > 40$, and $f'(x) < 0$ if $x < 40$, so f is minimized when $x = 40$. The minimum length of fencing required for this field is 240 feet.

85. Let x represent the length of each of the dividers. Then the field is x by A/x ft. We minimize the total length of fencing, given by:

$$f(x) = (n+2)x + \frac{2A}{x}, \qquad 0 < x < \infty.$$

Now $f'(x) = n + 2 - \frac{2A}{x^2}$, which is zero only when $x = \sqrt{\dfrac{2A}{n+2}}$. Verification: $f'(x) > 0$ if $x > \sqrt{\dfrac{2A}{n+2}}$ and $f'(x) < 0$ if $x < \sqrt{\dfrac{2A}{n+2}}$, so f is minimized when $x = \sqrt{\dfrac{2A}{n+2}}$. The minimum length of fencing required for this field is

$$f\left(\sqrt{\frac{2A}{n+2}}\right) = (n+2)\sqrt{\frac{2A}{n+2}} + \frac{2A\sqrt{n+2}}{\sqrt{2A}}$$

$$= \sqrt{2A(n+2)} + \sqrt{2A(n+2)}$$

$$= 2\sqrt{2A(n+2)} \text{ (ft)}.$$

88. Let L be the line segment in the first quadrant that is tangent to the graph of $y = 1/x$ at $(x, 1/x)$ and has endpoints $(0, c)$ and $b, 0)$. Compute the slope of L in several ways: as the value of dy/dx at the point of tangency, as the slope of the line segment between $(x, 1/x)$ and $(b, 0)$, and as the slope of the line segment between $(x, 1/x)$ and $(0, c)$:

$$-\frac{1}{x^2} = \frac{\frac{1}{x} - 0}{x - b} \text{ so } b = 2x.$$

$$-\frac{1}{x^2} = \frac{\frac{1}{x} - c}{x} : c - \frac{1}{x} = \frac{1}{x}, \text{ so } c = \frac{2}{x}.$$

Therefore the area A of the triangle is $A = A(x) = \dfrac{1}{2} \cdot 2x \cdot \dfrac{1}{x} = 1$. Because A is a constant function, every triangle has both maximal and minimal area.

91. Let x be the width of the box. Then its length is $5x$, and its height is $225/\left(5x^2\right)$. We minimize its surface area A, where

$$A(x) = 10x^2 + \frac{225}{5x^2} \cdot 2(6)x = 10x^2 + \frac{(12)(225)}{5x} \qquad 0 < x < \infty.$$

Now $A'(x) = 20x - \dfrac{(12)(225)}{5x^2}$, and $A'(x) = 0$ when $x = \sqrt[3]{(6)(225)/50}$. Verification: $A'(x) > 0$ for $x > \sqrt[3]{(6)(225)/50}$, and $A'(x) < 0$ for $x < \sqrt[3]{(6)(225)/50}$, so at this critical point the surface area is minimized. The minimal surface area is 270 cm^2.

94. Let θ be the angle between your initial path and due north, so that $0 \le \theta \le \pi/2$, and if $\theta = \pi/2$ then you plan to jog around a semicircle and not swim at all. Suppose that you can swim with speed v (in miles per hour). Then you will swim a length of $2\cos\theta$ miles at speed v and jog a length of 2θ miles at speed $2v$, for a total time of

$$T(\theta) = \frac{2\cos\theta}{v} + \frac{2\theta}{2v} = \frac{1}{v}(\theta + 2\cos\theta), \qquad 0 \le \theta \le \pi/2.$$

It's easy to verify that this formula is correct even in the extreme case $\theta = \pi/2$. It turns out that although $T'(\theta) = 0$ when $\theta = \pi/6$, this value of θ actually *maximizes* $T(\theta)$; this function has an endpoint minimum not even at $\theta = 0$, but at $\theta = \pi/2$. Answer: Jog all the way.

Chapter 5: The Integral

Section 5.2

4. $\int \left(-\frac{1}{t^2}\right) dt = \frac{1}{t} + C$

10. $\int \left(2x\sqrt{x} - \frac{1}{\sqrt{x}}\right) dx = \frac{4}{5}x^{5/2} - 2x^{1/2} + C$

16. $\int (t+1)^{10} \, dt = \frac{1}{11}(t+1)^{11} + C$

22. $\int \frac{(3x+4)^2}{\sqrt{x}} \, dx = \int x^{-1/2}\left(9x^2 + 24x + 16\right) dx$

$$= \int \left(9x^{3/2} + 24x^{1/2} + 16x^{-1/2}\right) dx = \frac{18}{5}x^{5/2} + 16x^{3/2} + 32x^{1/2} + C$$

28. $\int (2\cos \pi x + 3\sin \pi x) \, dx = \frac{2}{\pi}\sin \pi x - \frac{3}{\pi}\cos \pi x + C$

31. $\frac{1}{2}\sin^2 x + C_1 = -\frac{1}{2}\cos^2 x + C_2$: $\sin^2 x + \cos^2 x + 2C_1 = 2C_2$; $1 + 2C_1 = 2C_2$; $C_2 = C_1 + \frac{1}{2}$.

34. **(a)** $D_x \tan x = \sec^2 x$; **(b)** $\int \tan^2 x \, dx = \int (\sec^2 x - 1) \, dx = \tan x - x + C.$

40. $y = \int \sqrt{x+9} \, dx = \frac{2}{3}(x+9)^{3/2} + C$; $0 = y(-4) = \frac{2}{3}(-4+9)^{3/2} + C = \frac{2}{3}\cdot 5\sqrt{5} + C$;

$y(x) = \frac{2}{3}(x+9)^{3/2} - \frac{10}{3}\sqrt{5}.$

46. $y = \int (2x+3)^{3/2} \, dx = \frac{1}{5}(2x+3)^{5/2} + C = \frac{1}{5}(2x+3)^{5/2} + \frac{257}{5}$

52. $x(t) = 4t - 2\cos 2t$

55. The graph is shown below.

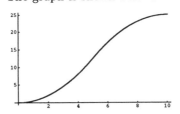

56. The graph is shown below.

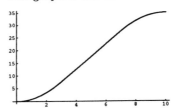

In the solutions for Problems 57–78, unless otherwise indicated, we will take the upward direction to be the positive direction, $s = s(t)$ for position (in feet) at time t (in seconds) with $s = 0$ corresponding to ground level, and $v(t)$ velocity at time t in ft/s, $a = a(t)$ acceleration at time t in ft/s². The initial position will be denoted by s_0 and the initial velocity by v_0.

58. With initial velocity v_0, here we have

$$a(t) = -32, \quad v(t) = -32t + v_0, \quad \text{and} \quad s(t) = -16t^2 + v_0 t$$

(because $s_0 = 0$). The maximum altitude is attained when $v = 0$, which occurs when $t = v_0/32$. Therefore

$$400 = s(v_0/32) = (-16)(v_0/32)^2 + (v_0)^2/32.$$

It follows that $\frac{1}{64}(v_0)^2 = 400$, and therefore that $v_0 = 160$ (ft/s).

61. Here, $v(t) = -32t + 48$ and $s(t) = -16t^2 + 48 + 160$. The ball strikes the ground at that value of $t > 0$ for which $s(t) = 0$:

$$0 = s(t) = -16(t-5)(t+2), \text{ so } t = 5.$$

Therefore the ball remains aloft for 5 seconds. Its velocity at impact is $v(5) = -112$ (ft/s), so the ball strikes the ground at a speed of 112 ft/s.

64. Here we have $v(t) = -32t + 320$ and $s(t) = -16t^2 + 320t$. After three seconds have elapsed, the height of the arrow will be $s(3) = 816$ (ft). The height of the arrow will be 1200 feet when $s(t) = 1200$:

$$16t^2 - 320t + 1200 = 0;$$
$$16(t - 5)(t - 15) = 0;$$

$t = 5$ and $t = 15$ are both solutions. So the height of the arrow will be 1200 feet both at $t = 5$ (the arrow is still rising) and at $t = 15$ (the arrow is falling). The arrow strikes the ground at that value of $t > 0$ for which $s(t) = 0$: $t = 20$. So the arrow will strike the ground 20 seconds after it is released.

67. In this problem, $v(t) = -32t + v_0 = -32t$ and $s(t) = -16t^2 + s_0 = -16t^2 + 400$. The ball reaches the ground when $s = 0$, thus when $16t^2 = 400$: $t = +5$. Ths impact velocity is $v(5) = (-32)(5) = -160$ (ft/s).

70. Let $f(t)$ be the altitude of the first ball at time t. Then $f(t) = -16t^2 + h$, so the altitude of the first ball will be $h/2$ when $f(t) = h/2$:

$$t^2 = \frac{h}{32}, \text{ so } t = \frac{1}{8}\sqrt{2h}.$$

Let the second ball have initial velocity v_0 and altitude $s(t)$ at time t. Then $s(t) = -16t^2 + v_0 t$. We require that $s(t) = h/2$ at the above value of t. That is,

$$\frac{h}{2} = (-16)\left(\frac{h}{2}\right) + (v_0)\left(\frac{1}{8}\sqrt{2h}\right).$$

Solution of this equation yields $v_0 = 4\sqrt{2h}$.

73. Bomb equations: $a = -32$, $v = -32t$, $s_B = s = -16t^2 + 800$. Here we have $t = 0$ at the time the bomb is released.
Projectile equations: $a = -32$, $v = -32(t - 2) + v_0$, and $s_P = s = -16(t - 2)$, $t \geq 2$.
We require $s_B = s_P = 400$ at the same time. The equation $s_B = 400$ leads to $t = 5$, and for $s_P(5) = 400$, we must have $v_0 = 544/3 \approx 181.33$ (ft/s).

76. Let $x(t)$ be the altitude (in miles) of the spacecraft at time t (hours), with $t = 0$ corresponding to the time at which the retrorockets are fired; let $v(t) = x'(t)$ be the velocity of the spacecraft at time t. Then $v_0 = -1000$ and x_0 is unknown. But the acceleration constant is $a = +20000$, so

$$v(t) = (20000)t - 1000 \text{ and } x(t) = (10000)t^2 - 1000t + x_0.$$

We want $v = 0$ exactly when $x = 0$—call the time then t_1. Then $0 = (20000)t_1 - 1000$, so $t_1 = 1/20$. Also $x(t_1) = 0$, so

$$0 = (10000)\left(\frac{1}{400}\right) - (1000)\left(\frac{1}{20}\right) + x_0.$$

Therefore $x_0 = 50 - 25 = 25$ miles. (Also $t_1 = 1/20$ of an hour; that is, exactly three minutes.)

Section 5.3

1. $\displaystyle\sum_{i=1}^{5} 3^i = 3^1 + 3^2 + 3^3 + 3^4 + 3^5 = 3 + 9 + 27 + 81 + 243.$

4. $1 + 3 + 5 + 7 + 9 + 11$

7. $x + x^2 + x^3 + x^4 + x^5$

10. $\displaystyle\sum_{n=1}^{6} n \cdot (-1)^{n+1}$

13. $\displaystyle\sum_{m=1}^{6} \frac{1}{2^m}$

16. $\displaystyle\sum_{j=1}^{9} \sqrt{j}$

22. -231

25. $\displaystyle\sum_{i=1}^{6} (i^3 - i^2) = \frac{6^2 \cdot 7^2}{4} - \frac{6 \cdot 7 \cdot 13}{6} = 3^2 \cdot 7^2 - 7 \cdot 13 = 441 - 91 = 350.$

28. $\displaystyle\frac{10^8}{4} + \frac{10^6}{2} + \frac{10^4}{4} = 25{,}502{,}500.$

31. n^2

34. $\displaystyle \underline{A}_5 = \sum_{i=1}^{5} \frac{2i+3}{5} \cdot \frac{1}{5}, \qquad \overline{A}_5 = \sum_{i=1}^{5} \frac{2i+5}{5} \cdot \frac{1}{5}.$

37. $\displaystyle \underline{A}_5 = \sum_{i=1}^{5} \left(\frac{i-1}{5}\right)^2 \cdot \frac{1}{5}, \qquad \overline{A}_5 = \sum_{i=1}^{5} \left(\frac{i}{5}\right)^2 \cdot \frac{1}{5}.$

40. $\displaystyle \underline{A}_8 = \sum_{i=1}^{8} \left[9 - \left(\frac{i}{4}+1\right)^2 \right] \cdot \frac{1}{8}, \qquad \overline{A}_8 = \sum_{i=1}^{8} \left[9 - \left(\frac{i-1}{4}+1\right)^2 \right] \cdot \frac{1}{8}.$

46. $\displaystyle \sum_{i=1}^{n} \left(\frac{2i}{n^2}\right)^2 \cdot \frac{2}{n} = \frac{8n(n+1)(2n+1)}{6n^3} \to \frac{8}{3}$ as $n \to \infty.$

49. $\displaystyle \sum_{i=1}^{n} \left(5 - \frac{3i}{n}\right) \left(\frac{1}{n}\right) = 5n \cdot \frac{1}{n} - \frac{3n(n+1)}{2n^2} \to \frac{7}{2}$ as $n \to \infty.$

Section 5.4

1. $\displaystyle\int_{1}^{3} (2x - 1)\, dx$

4. $\displaystyle\int_{0}^{3} (x^3 - 3x^2 + 1)\, dx$

7. $\displaystyle\int_{3}^{8} \frac{1}{\sqrt{1+x}}\, dx$

10. $\displaystyle\int_{0}^{\pi/4} \tan x\, dx$

13. $\displaystyle\sum_{i=1}^{n} f(x_i^*) \Delta x = \sum_{i=1}^{5} \frac{1}{1+i} = 1.45.$

16. $\displaystyle\sum_{i=1}^{n} f(x_i^*) \Delta x = \sum_{i=1}^{6} \left((1 + i/2)^2 + 2(1 + i/2)\right)(1/2) = 41.375$

19. $\displaystyle\sum_{i=1}^{6} \left(\cos \frac{i\pi}{6}\right) \cdot \frac{\pi}{6} = -\pi/6.$

22. 0.16

28. 4.092967 (rounded)

31. 33/100

34. 7.505140

40. $\displaystyle\sum_{i=1}^{6}\left(\sin\pi\cdot\frac{2i-1}{12}\right)\cdot\frac{1}{6}=\frac{\sqrt{2}+\sqrt{6}}{6}.$

43. $\displaystyle\sum_{i=1}^{n}\left(\frac{2i}{n}\right)^2\cdot\frac{2}{n}=\frac{8n(n+1)(2n+1)}{6n^3}\to\frac{8}{3}$ as $n\to\infty.$

46. $\displaystyle\sum_{i=1}^{n}\left[4-3\left(1+\frac{4i}{n}\right)\right]\cdot\frac{4}{n}=\frac{16}{n}\cdot n-\frac{12}{n}\cdot n-\frac{48n(n+1)}{2n^2}\to 4-24=-20$ as $n\to\infty.$

49. Take $x_i=\dfrac{bi}{n}$ and $\Delta x=\dfrac{b}{n}$. The integral is equal to $\displaystyle\lim_{n\to\infty}\frac{n(n+1)}{2n^2}b^2=\frac{1}{2}b^2.$

Section 5.5

1. $\displaystyle\int_0^1\left(3x^2+2\sqrt{x}+3\sqrt[3]{x}\right)dx=\left[x^3+\tfrac{4}{3}x^{3/2}+\tfrac{9}{4}x^{4/3}\right]_0^1=\frac{55}{12}.$

4. $\displaystyle\int_{-2}^{-1}\frac{1}{x^4}\,dx=\left[-\frac{1}{3x^3}\right]_{-2}^{-1}=\frac{7}{24}.$

7. $\displaystyle\int_{-1}^{0}(x+1)^3\,dx=\left[\tfrac{1}{4}(x+1)^4\right]_{-1}^{0}=\tfrac{1}{4}.$

10. $\displaystyle\int_{1}^{4}\frac{1}{\sqrt{x}}\,dx=\left[2\sqrt{x}\right]_1^4=2\sqrt{4}-2\sqrt{1}=2.$

13. $\displaystyle\int_{-1}^{1}x^{99}\,dx=\left[\tfrac{1}{100}x^{100}\right]_{-1}^{1}=0.$

16. $\displaystyle\int_{1}^{2}(x^2+1)^3\,dx=\int_{1}^{2}(x^6+3x^4+3x^2+1)\,dx=\left[\tfrac{1}{7}x^7+\tfrac{3}{5}x^5+x^3+x\right]_1^2=\frac{1566}{35}\approx 44.742857.$

19. $\displaystyle\int_{1}^{8}x^{2/3}\,dx=\left[\tfrac{3}{5}x^{5/3}\right]_1^8=\frac{93}{5}.$

22. $\displaystyle\int_{0}^{4}\sqrt{3t}\,dt=\left[\tfrac{2}{3}t^{3/2}\sqrt{3}\right]_0^4=\tfrac{16}{3}\sqrt{3}.$

25. $\displaystyle\int_{1}^{4}\frac{x^2-1}{\sqrt{x}}\,dx=\int_{1}^{4}(x^{3/2}-x^{-1/2})\,dx=\left[\tfrac{2}{5}x^{5/2}-2x^{1/2}\right]_1^4=\frac{52}{5}.$

28. $\displaystyle\int_{0}^{\pi/2}\cos 2x\,dx=\left[\tfrac{1}{2}\sin 2x\right]_0^{\pi/2}=0.$

31. $\displaystyle\int_{0}^{\pi}\sin 5x\,dx=\left[-\tfrac{1}{5}\cos x\right]_0^{\pi}=\tfrac{2}{5}.$

34. $\displaystyle\int_{0}^{5}\sin\frac{\pi x}{10}\,dx=\left[-\frac{10}{\pi}\cos\frac{\pi x}{10}\right]_0^{5}=\frac{10}{\pi}.$

37. Choose $x_i=i/n$, $\Delta x=1/n$, $x_0=0$, and $x_n=1$. Then the limit in question is the limit of a Riemann sum for the function $f(x)=2x-1$ on the interval $0\le x\le 1$, and its value is therefore

$$\int_0^1(2x-1)\,dx=\left[x^2-x\right]_0^1=1-1=0.$$

40. This limit is the integral of $f(x)=x^3$ on the interval $0\le x\le 1$, and it is therefore equal to $\tfrac{1}{4}$.

43. $\displaystyle\int_{-2}^{2} |1 - x|\, dx = 5.$

The graph is shown below.

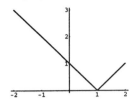

46. $\displaystyle\int_{0}^{6} |5 - |2x||\, dx = \frac{37}{2} = 18.5.$

The graph is shown below.

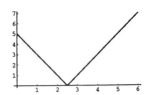

49. $0 \le x^2 \le x$ if $0 \le x \le 1$. Hence $1 \le 1 + x^2 \le 1 + x$ for such x. Therefore

$$1 \le \sqrt{1 + x^2} \le \sqrt{1 + x}$$

if $0 \le x \le 1$. Hence, by the comparison property, the inequality in Problem 49 follows.

52. $x^2 \le x^5$ if $x \ge 1$. So $1 + x^2 \le 1 + x^5$ if $2 \le x \le 5$. Therefore

$$\frac{1}{1 + x^5} \le \frac{1}{1 + x^2}$$

if $2 \le x \le 5$. The inequality in Problem 52 now follows from the comparison property.

55. If $0 \le x \le 1$, then

$$1 \le 1 + x \le 2;$$
$$\frac{1}{2} \le \frac{1}{1 + x} \le 1;$$
$$\frac{1}{2} \cdot (1 - 0) \le \int_{0}^{1} \frac{1}{1 + x}\, dx \le 1 \cdot (1 - 0).$$

So the value of the integral lies between 0.5 and 1.0.

58. If $0 \le x \le \frac{1}{4}\pi$, then

$$0 \le \sin x \le \frac{\sqrt{2}}{2};$$
$$0 \le \sin^2 x \le \frac{1}{2};$$
$$0 \le 2\sin^2 x \le 1;$$
$$16 \le 16 + 2\sin^2 x \le 17;$$
$$4 \le \sqrt{16 + 2\sin^2 x} \le \sqrt{17};$$
$$4 \cdot \frac{\pi}{4} \le \int_{0}^{\pi/4} \sqrt{16 + 2\sin^2 x}\, dx \le \frac{\pi\sqrt{17}}{4}.$$

Therefore

$$3.14159 \le \int_{0}^{\pi/4} \sqrt{16 + 2\sin^2 x}\, dx \le 3.2384.$$

64. In 1990 the population in thousands was

$$125 + \int_{t=0}^{20} (8 + (-0.5)t + (0.03)t^2)\, dt = 125 + \left[8t + (0.25)t^2 + (0.01)t^3 \right]_{0}^{20}$$
$$= 125 + 160 + 100 + 80 = 465 \text{ (thousands)}.$$

Section 5.6

1. $\dfrac{1}{2}\displaystyle\int_0^2 x^4\,dx = 16/5.$

4. $\dfrac{1}{4}\displaystyle\int_0^4 8x\,dx = 16.$

10. $\dfrac{2}{\pi}\displaystyle\int_0^{\pi/2} \sin 2x\,dx = \dfrac{2}{\pi}.$

16. $\displaystyle\int_{-1}^1 (x^3+2)^2\,dx = \left[\tfrac{1}{7}x^7 + x^4 + 4x\right]_{-1}^1 = \tfrac{58}{7}.$

22. On the interval $-1 \le x \le 1$ of integration, the integrand is equal to $x-2$, so the value of the integral is -4.

25. Split the integral into three integrals—one on the interval from -2 to -1, one on the interval from -1 to 1, and one on the interval from 1 to 2.

28. $\displaystyle\int_5^{10} \dfrac{dx}{\sqrt{x-1}} = \left[2\sqrt{x-1}\right]_5^{10} = 2.$

31. $\displaystyle\int_{-3}^0 (x^3 - 9x)\,dx - \int_0^3 (x^3 - 9x)\,dx = \left[\tfrac{1}{4}x^4 - \tfrac{9}{2}x^2\right]_{-3}^0 - \left[\tfrac{1}{4}x^4 - \tfrac{9}{2}x^2\right]_0^3 = \tfrac{81}{4} + \tfrac{81}{4} = \tfrac{81}{2}.$

34. $P_{\text{AV}} = \dfrac{1}{10}\displaystyle\int_0^{10} \left(100 + 10t + (0.02)t^2\right)\,dt = \dfrac{1}{10}\left[100t + 5t^2 + \dfrac{0.02}{3}t^3\right]_0^{10}$

$\qquad = \tfrac{1}{10}\left(1000 + 500 + \tfrac{20}{3}\right) = \tfrac{452}{3} \approx 150.667.$

37. $T_{\text{AV}} = \dfrac{1}{10}\displaystyle\int_0^{10} (40x - 4x^2)\,dx = \dfrac{1}{10}\left[20x^2 - \dfrac{4}{3}x^3\right]_0^{10} = \dfrac{1}{10}\left(2000 - \dfrac{4000}{3}\right) = \dfrac{200}{3}.$

40. Let $v(t)$ represent velocity of the car. Then $v(t) = at$ and the position of the car at time t is $x(t) = \tfrac{1}{2}at^2$. So the final velocity, at time $t = T$, is $v(T) = aT$. Average velocity during the T seconds:

$$v_{\text{AV}} = \dfrac{1}{T}\int_0^T at\,dt = \dfrac{1}{T}\left[\dfrac{1}{2}at^2\right]_0^T = \dfrac{1}{2}aT.$$

The final position, at time $t = T$, is $x(T) = \tfrac{1}{2}aT^2$. Average position during the T seconds:

$$x_{\text{AV}} = \dfrac{1}{T}\int_0^T \dfrac{1}{2}at^2\,dt = \dfrac{1}{T}\cdot\dfrac{a}{2}\cdot\left[\dfrac{1}{3}t^3\right]_0^T = \dfrac{a}{6T}\cdot T^3 = \dfrac{1}{6}aT^2.$$

43. Part (a): $A(x) = 2x\sqrt{16-x^2}$, $0 \le x \le 4$. Part (b):

$$\dfrac{1}{4}\int_0^4 A(x)\,dx = \dfrac{32}{3}.$$

Part (c): $A(x) = \tfrac{32}{3}$ for two values of x in $[0,4]$. Thus there are two such rectangles. One has base of length $\tfrac{4}{3}\sqrt{18 + 6\sqrt{5}} \approx 7.473379$ and the other has base of length $\tfrac{4}{3}\sqrt{18 - 6\sqrt{5}} \approx 2.854577.$

46. $g'(t) = \left(t^2 + 25\right)^{1/2}$

49. $f'(x) = -x - \dfrac{1}{x}$

52. $G'(x) = \sin^3 x$

58. $f'(x) = \left(1 + \sin^2 x\right)^3 \cos x$

61. $dy = \dfrac{1}{x}\,dx$: $y(x) = \displaystyle\int_1^x \dfrac{1}{t}\,dt.$

64. $y(x) = 2 + \displaystyle\int_1^x \tan t\,dt.$

67. Part (a): $g(0) = 0$, $g(2) = 4$, $g(4) = 8$, $g(6) = 4$, $g(8) = -4$, and $g(10) = -8$. Part (b): The function g is increasing on $(0, 4)$ and decreasing on $(4, 10)$. Part (c): Its maximum value is 8 and its minimum value is -8. Part (d): The graph of $g(x)$ is shown on the right.

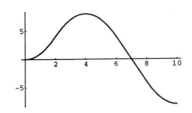

70. Part (a): The function g has local extrema when $x = 0$, π, 2π, 3π, and 4π. Part (b): The global minimum of g is $g(0) = 0$ and the global maximum occurs where $x = \pi$. Part (c): The extrema of f correspond to inflection points of g. The approximate x-coordinates of these points are 4.49493409, 7.725252, and 10.904122. Part (d): The graph of $g(x)$ is shown on the right.

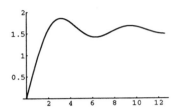

Section 5.7

1. Let $u = 3x - 5$. Then $du = 3\,dx$, so $dx = \frac{1}{3}\,du$. Thus

$$\int (3x - 5)^{17}\,dx = \int \frac{1}{3}u^{17}\,du = \frac{1}{54}u^{18} + C = \frac{1}{54}(3x - 5)^{18} + C.$$

4. The given substitution yields

$$\int \frac{1}{6}u^{-1/3}\,du = \frac{1}{4}u^{2/3} + C = \frac{1}{4}(2x^3 - 1)^{2/3} + C.$$

7. The given substitution yields

$$\int \frac{1}{4}\sin u\,du = -\frac{1}{4}\cos u + C = -\frac{1}{4}\cos(2x^2) + C.$$

10. The given substitution yields

$$\int \frac{1}{6}u^{-1/2}\,du = \frac{1}{3}u^{1/2} + C = \frac{1}{3}\sqrt{5 + 2\sin 3x} + C.$$

16. $\displaystyle\int \frac{dx}{(3 - 5x)^2} = \frac{1}{5(3 - 5x)} + C$

19. $\displaystyle\int \sec 2\theta \tan 2\theta\,d\theta = \frac{1}{2}\sec 2\theta + C$

22. $\displaystyle\int 3t\left(1 - 2t^2\right)^{10}\,dt = -\frac{3}{44}\left(1 - 2t^2\right)^{11} + C$

28. $\displaystyle\int t\sec^2 t^2\,dt = \frac{1}{2}\tan t^2 + C$

34. $\displaystyle\int \sec^3 \theta \tan \theta\,d\theta = \frac{1}{3}\sec^3 \theta + C$

40. $\displaystyle\int \frac{2 - x^2}{\left(x^3 - 6x + 1\right)^5}\,dx = \frac{1}{12}\left(x^3 - 6x + 1\right)^{-4} + C$

46. $\displaystyle\int_0^{\pi/2} \sin x \cos x \, dx = \left[\frac{1}{2}\sin^2 x\right]_0^{\pi/2} = \frac{1}{2}$

52. $\displaystyle\int_0^{\pi/2} (\cos x)\sqrt{\sin x} \, dx = \left[\frac{2}{3}(\sin x)^{3/2}\right]_0^{\pi/2} = \frac{2}{3}$

58. $\displaystyle\int_0^1 \cos^2 \pi t \, dt = \left[\frac{1}{2}t + \frac{\sin 2\pi t}{4\pi}\right]_0^1 = \frac{1}{2}$

64. $\frac{1}{2}\sec^2\theta + C_2 = \frac{1}{2}(1 + \tan^2\theta) + C_2 = \frac{1}{2}\tan^2\theta + \frac{1}{2} + C_2$, so $C_1 = \frac{1}{2} + C_2$

67. $\displaystyle\int_{-a}^a f(x)\,dx = \int_{-a}^0 f(x)\,dx + \int_0^a f(x)\,dx = -\int_a^0 f(-u)\,du + \int_0^a f(x)\,dx$

$\displaystyle = \int_0^a f(-u)\,du + \int_0^a f(x)\,dx = -\int_0^a f(u)\,du + \int_0^a f(x)\,dx = 0$

Section 5.8

1. $\displaystyle\int_{-4}^4 (16 - x^2)\,dx = \frac{256}{3}.$

4. $\displaystyle\int_0^3 (9x - x^3)\,dx = \frac{81}{4}.$

7. $\displaystyle\int_{-2}^2 (16 - 4x^2)\,dx = \frac{128}{3}.$

10. $\displaystyle\int_0^4 (4x - x^2)\,dx = \frac{32}{3}.$

16. $A = \displaystyle\int_0^4 (4x - x^2)\,dx = \left[2x^2 - \frac{1}{3}x^3\right]_0^4 = \frac{32}{3}.$

22. $\displaystyle\int_{-2}^2 (8 - 2x^2)\,dx = \frac{64}{3}.$

25. $\displaystyle\int_{-1}^3 (2x + 3 - x^2)\,dx = \frac{32}{3}.$

28. $\displaystyle\int_{-4}^2 (8 - 2y - y^2)\,dy = 36.$

31. $\displaystyle\int_1^{25} \left(\frac{1}{12}(-5 - x) + \frac{1}{2}\sqrt{x}\right)dx = \frac{16}{3}.$

34. $\displaystyle\int_{2/3}^2 (x^2 - 4(x - 1)^2)\,dx = \frac{32}{27}.$

37. $\displaystyle\int_0^4 (32\sqrt{x} - x^3)\,dx = \frac{320}{3}.$

40. $\displaystyle\int_{-\sqrt{6}}^{\sqrt{6}} \left(\frac{1}{2}y^2 + 3 - y^2\right)dy = 4\sqrt{6}.$

43. $\displaystyle\int_{-1}^0 (g(x) - f(x))\,dx + \int_0^2 (f(x) - g(x))\,dx = \frac{37}{12}.$

46. If we let $u = x^2$, then $dy = 2x\,dx$ and the new limits of integration will be 0 and 9. The resulting integral is

$$\int_0^9 \frac{1}{2}\sqrt{81 - u^2}\,du,$$

which is one-eighth the area of a circle of radius 9. Hence the value of the integral is $\frac{81}{8}\pi$.

49. The points of intersection are $A(-1,1)$ and $B(2,4)$, and C is the point $C(1/2, 1/4)$. The normal line at C has the equation $4x + 4y = 3$. The line through A and B has equation $y = x + 2$. These lines meet at $(-5/8, 11/8)$. The altitude of the triangle ABC from its vertex at C has length $\frac{9}{8}\sqrt{2}$ and the base of the triangle has length $3\sqrt{2}$. So the area of the triangle is $\frac{27}{8}$. The area of the segment is $\frac{9}{2}$, and this is indeed four-thirds the area of the triangle.

52. First, it is not difficult to show that the integral of $f(x)$ over the interval $-h \le x \le h$ is $\frac{2}{3}ph^3 + 2rh$. But with $a = -h$, $m = 0$, and $b = h$, we find that

$$\frac{h}{3}[f(a) + 4f(m) + f(b)] = \frac{2}{3}ph^2 + 2rh$$

as well. Therefore for $a = -h$, $m = 0$, and $b = h$, we have

$$\int_a^b f(x)\,dx = \frac{h}{3}[f(a) + 4f(m) + f(b)].$$

But the right-hand side above depends only on $f(a)$, $f(m)$, $f(b)$, and h, so it is independent of translation along the x-axis. The result in Problem 52 now follows.

55. The curves cross where $x = a \approx -0.824132312$ and where $x = b \approx 0.8324132312$, and the area between them is

$$\int_a^b \left[(\cos x) - x^2\right]\,dx \approx 1.09475.$$

58. The curves cross where $x = a \approx -1.752171779$ and where $x = 2$, and the area between them is

$$\int_a^2 \left(2x - x^2 - x^4 + 16\right)\,dx \approx 46.801821.$$

Section 5.9

1. $T_4 = 8$, and the true value of the integral is also 8.

4. $T_4 = 0.71$, and the true value of the integral is $\frac{2}{3}$.

7. $M_4 = 8$.

10. $M_4 = 0.65$.

13. $T_4 = 8.75$, $S_4 \approx 8.6667$, integral: $\frac{26}{3}$.

16. $T_4 \approx 1.2182$, $S_4 \approx 1.2189$, integral: $\frac{2}{3}(2\sqrt{2} - 1) \approx 1.218951417$.

19. $T_8 \approx 8.5499$, $S_8 \approx 8.5509$, integral: approximately 8.550733044.

22. (a) 28.3;   (b) 28.6.

28. $f^{(4)}(x) = \dfrac{24}{x^5}$, and the maximum of $\left|f^{(4)}(x)\right|$ for $1 \le x \le 2$ is 24. We take $M = 24$, $a = 1$, $b = 2$, and ask for n large enough that

$$\frac{M(b-a)^5}{180n^4} < 0.000005;$$

thus $n^4 > \dfrac{2}{(15)(0.000005)}$, and therefore $n \ge 14$.

Chapter 5 Miscellaneous

1. $\displaystyle\int \frac{x^5 - 2x + 5}{x^3}\, dx = \frac{1}{3}x^3 + \frac{2}{x} - \frac{5}{2x^2} + C$

4. $\displaystyle\int 7(2x+3)^{-3}\, dx = -\frac{7}{4}(2x+3)^{-2} + C$

10. $\displaystyle\int 3x\left(1+3x^2\right)^{-1/2}\, dx = \sqrt{1+3x^2} + C$

16. $\displaystyle\int \left(1+x^{1/2}\right)^{-2} x^{-1/2}\, dx = \int (1+u)^{-2}\, 2\, du = -\frac{2}{1+u} + C = -\frac{2}{1+\sqrt{x}} + C$

22. $\displaystyle\int \frac{x + 2x^3}{\left(x^4 + x^2\right)^3}\, dx = \int \frac{1}{2}u^{-3}\, du = -\frac{1}{4}\left(x^4 + x^2\right)^{-2} + C$

28. $\displaystyle y(x) = \int \frac{2}{\sqrt{x+5}}\, dx = 4(x+5)^{1/2} + C; \quad y(4) = 3 = 12 + C$, so $C = -9$; $y(x) = 4\sqrt{x+5} - 9$.

34. With the usual notation, $a = +8$, $v = 8t$, and $s = 4t^2$. When $v = 88$, $t = 11$, so the distance traveled then will be $(4)(11)^2 = 484$ feet.

37. Assume that the brakes are applied at time $t = 0$ (seconds), that $s(t)$ is the distance the car subsequently travels, that $v(t)$ is its velocity at time t, and that its constant deceleration is $-a$ (where $a > 0$). Then

$$s = -\frac{1}{2}at^2 + v_0 t \text{ and } v = -at + v_0.$$

Now $v = 0$ when $t = v_0/a$, and it follows from the first equation above that $a = 22$. So $s = -11t^2 + v_0 t$ and $v = -22t + v_0$. But $v = 0$ when $t = 4$, and at this time we have $s = 176$ feet. The point of this problem is that *doubling* the speed *quadruples* the stopping distance.

40. $\displaystyle\sum_{k=1}^{100}\left(\frac{1}{k} - \frac{1}{k+1}\right) = 1 - \frac{1}{2} + \frac{1}{2} - \frac{1}{3} + \frac{1}{3} - \frac{1}{4} + \cdots + \frac{1}{98} - \frac{1}{99} + \frac{1}{99} - \frac{1}{100} + \frac{1}{100} - \frac{1}{101} = 1 - \frac{1}{101} = \frac{100}{101}.$

43. $\displaystyle\lim_{n\to\infty}\sum_{i=1}^{n}\frac{\Delta x}{\sqrt{x_i^*}} = \int_1^2 x^{-1/2}\, dx = \left[2\sqrt{x}\right]_1^2 = 2\left(\sqrt{2} - 1\right).$

46. $\displaystyle\lim_{n\to\infty}\frac{1^{10} + 2^{10} + 3^{10} + \cdots + n^{10}}{n^{11}} = \int_0^1 x^{10}\, dx = \left[\frac{1}{11}x^{11}\right]_0^1 = \frac{1}{11}.$

52. $\displaystyle\int \frac{\left(1+\sqrt[3]{x}\right)^2}{\sqrt{x}}\, dx = \int \left(x^{-1/2} + 2x^{-1/6} + x^{1/6}\right)\, dx = 2x^{1/2} + \frac{12}{5}x^{5/6} + \frac{6}{7}x^{7/6} + C.$

55. You may use the substitution $u = x^{3/2}$, $du = \frac{3}{2}x^{1/2}$.

58. $\displaystyle\int_1^2 \frac{2t+1}{\sqrt{t^2+t}}\, dt = \left[2(t^2+t)^{1/2}\right]_1^2 = 2\sqrt{2}\left(\sqrt{3} - 1\right)$ (using the substitution $u = t^2 + t$).

61. One easy substitution is $u = 1 + \sqrt{t}$.

64. $\displaystyle A = \int_{-1}^1 \left(1 - x^3\right)\, dx = \left[x - \frac{1}{4}x^4\right]_{-1}^1 = 2.$

67. The curves cross at $(-1, 1)$ and at $(1, 1)$; we obtain the total area by doubling the integral of $2 - x^2 - x^4$ over the interval $0 \le x \le 1$.

70. The curves cross at $(-1, 1)$ and at $(1, 1)$. By symmetry,

$$A = 2\int_0^1 \left(2 - x^2 - x^{2/3}\right)\, dx = 2\left[2x - \frac{1}{3}x^3 - \frac{3}{5}x^{5/3}\right]_0^1 = \frac{32}{15}.$$

73. Differentiation of both sides of the given equation yields

$$2x = \left(1 + [f(x)]^2\right)^{1/2}, \text{ so that } 4x^2 = 1 + [f(x)]^2, \text{ and therefore } f(x) = \left(4x^2 - 1\right)^{1/2}.$$

This answer must be tested by substitution in the original equation because it was obtained under the assumption that such a function actually exists.

76. $T_6 \approx 2.812254$ and $S_6 \approx 2.828502$. For the exact value:

$$(1 - \cos x)^{1/2} = \sqrt{2} \left(\frac{1 - \cos x}{2} \right)^{1/2} = \sqrt{2} \left| \sin \frac{x}{2} \right|.$$

Therefore

$$\int_0^\pi (1 - \cos x)^{1/2} \, dx = \int_0^\pi \sqrt{2} \left(\frac{1 - \cos x}{2} \right)^{1/2} dx = \sqrt{2} \int_0^\pi \left| \sin \frac{x}{2} \right| dx = \left[-2\sqrt{2} \cos \frac{x}{2} \right]_0^\pi = 2\sqrt{2}.$$

Chapter 6: Applications of the Integral

Section 6.1

1. $\displaystyle \lim_{n\to\infty} \sum_{i=1}^{n} 2x_i^* \, \Delta x = \int_0^1 2x \, dx = 1.$

4. $\displaystyle \lim_{n\to\infty} \sum_{i=1}^{n} \left(3\left(x_i^*\right) - 1\right) \Delta x = \int_{-1}^3 \left(3x^2 - 1\right) dx = 24.$

7. $\displaystyle \lim_{n\to\infty} \sum_{i=1}^{n} (2m_i - 1) \Delta x = \int_{-1}^3 (2x - 1) \, dx = 4.$

10. $\displaystyle \lim_{n\to\infty} \sum_{i=1}^{n} m_i \cos\left(m_i\right)^2 \Delta x = \int_0^{\sqrt{\pi}} x \cos\left(x^2\right) dx = \left[\frac{1}{2} \sin\left(x^2\right)\right]_0^{\sqrt{\pi}} = 0.$

13. $\displaystyle \lim_{n\to\infty} \sum_{i=1}^{n} \sqrt{1 + [f\left(x_i^*\right)]^2} \, \Delta x = \int_0^{10} \left(1 + [f(x)]^2\right)^{1/2} dx$

16. $\displaystyle M = \int_0^{25} (60 - 2x) \, dx \, dx = \left[60x - x^2\right]_0^{25} = 875$

19. $\displaystyle \int_0^{10} -32t \, dt = -320$ is the net distance; $\displaystyle \int_0^{10} 32t \, dt = 320$ is the total distance.

22. $\displaystyle \int_0^5 |2t - 5| \, dt = 2 \int_0^{5/2} (5 - 2t) \, dt = \frac{25}{2}$ is the net and the total distance.

28. Net distance: 2. Total distance: $2\sqrt{2}$.

31. In this problem, $v(t) < 0$ on $(a, b) \approx (0.602750, 2.292402)$, so the total distance traveled is

$$\int_0^a v(t) \, dt - \int_a^b v(t) \, dt + \int_b^3 v(t) \, dt \approx 4.66407.$$

The net distance traveled is $\displaystyle \int_0^3 v(t) \, dt = \frac{3}{4}.$

34. In this problem, $v(t) < 0$ on $(a, b) \approx (2.167455, 5.128225)$. So the total distance traveled is

$$\int_0^a v(t) \, dt - \int_a^b v(t) \, dt + \int_b^{2\pi} v(t) \, dt \approx 7.68273.$$

The total distance traveled is $\displaystyle \int_0^{2\pi} v(t) \, dt \approx -0.430408.$

37. $\displaystyle M = \int_0^5 2\pi x \left(25 - x^2\right) dx = \left[25\pi x^2 - \frac{1}{2}\pi x^4\right]_0^5 = \frac{625}{2}\pi.$

40. $\displaystyle \int_0^{20} (13 + t) \, dt = 460$ (thousands of births).

43. $r(0) = 0.1 = a - b$, $r(182.5) = a + b$. So $a = 0.3$ and $b = 0.2$. Therefore the average total annual rainfall will be

$$R = \int_0^{365} \left(0.3 - 0.2\cos\frac{2\pi t}{365}\right) dt = (0.3)(365) - \frac{36.5}{\pi}\sin(2\pi) = 109.5 \text{ (inches)}.$$

Section 6.2

1. $V = \int_0^1 \pi x^4 \, dx = \dfrac{\pi}{5}.$

4. $V = \int_{0.1}^1 \dfrac{\pi}{x^2} \, dx = 9\pi.$

7. $V = \int_0^1 \left(\pi x - \pi x^4 \right) dx = \dfrac{3}{10}\pi.$

10. $V = \int_{-2}^3 \pi \left[(y+6)^2 - (y^2)^2 \right] dy = \dfrac{500}{3}\pi.$

13. $V = \int_0^1 \pi \left(\sqrt{1-y} \right)^2 dy = \dfrac{\pi}{2}.$

16. $V = \int_0^1 \pi \left[\left(2 + \sqrt{1-y} \right)^2 - \left(2 - \sqrt{1-y} \right)^2 \right] dy = \dfrac{16}{3}\pi.$

19. $V = \int_0^4 \pi \left(\sqrt{y} \right)^2 dy = 8\pi.$

22. $V = 2 \int_0^2 \pi \left[(9 - x^2)^2 - (x^2 + 1)^2 \right] dx = \dfrac{640}{3}\pi \approx 670.206433.$
Note how we used the fact that the plane region is symmetric around the y-axis.

25. The volume is $V = \int_0^\pi \pi \sin^2 x \, dx = \left[\frac{1}{2}\pi x - \frac{1}{4}\pi \sin 2x \right]_0^\pi = \frac{1}{2}\pi^2.$

28. The volume is

$$V = \int_{-\pi/3}^{\pi/3} \pi \left[(\cos^2 x) - \tfrac{1}{4} \right] dx = \tfrac{1}{4}\pi \left[x - \sin 2x \right]_{-\pi/3}^{\pi/3} = \tfrac{1}{12}\pi \left(2\pi + 3\sqrt{3} \right) \approx 3.005283590.$$

31. The curves intersect in three points, with x-coordinates $a \approx -0.532$, $b \approx 0.653$, and $c \approx 2.879$. Two bounded regions in the plane are formed by these curves. When the one on the left is rotated around the x-axis, the volume swept out is approximately 2.998317. When the one on the right is rotated around the x-axis, the volume swept out is approximately 267.441647. The total volume swept out by the two regions is approximately 270.439964.

34. The curves intersect in two points, with x-coordinates $a \approx 0.386236871$ and $b \approx 1.964192887$. When the region between them is rotated around the x-axis, the volume swept out is

$$V = \int_a^b \pi \left[(\sin^2 x) - (x-1)^4 \right] dx$$
$$= \pi \left[\tfrac{1}{10} \left(-2x^5 + 10x^4 - 20x^3 + 20x^2 - 5x \right) - \tfrac{1}{4}\sin 2x \right]_a^b \approx 3.00464.$$

37. Here, $x^2 = \dfrac{a^2}{b^2}(b^2 - y^2).$ So $V = 2 \int_0^b \pi \dfrac{a^2}{b^2}(b^2 - y^2) \, dy = \dfrac{4}{3}\pi a^2 b.$

40. Let AB be the segment on the x-axis from $x = -a$ to $x = a$. Visualize a cross section of the solid at x and perpendicular to AB. Its radius is then $y = (a^2 - x^2)^{1/2}$, and so it has area $\frac{1}{2}\pi y^2$. Therefore its volume is $V = 2 \int_0^a \dfrac{\pi}{2}(a^2 - x^2) \, dx = \dfrac{2}{3}\pi a^3.$

46. Set up a coordinate system with the center of the sphere at the origin and with the axis of the hole coincident with the y-axis. The volume that remains after the hole is drilled is then

$$V = 2 \int_0^4 \pi \left[(25 - y^2) - 9 \right] dy = \dfrac{256\pi}{3}.$$

52. (a) $x = (y/k)^{1/4}$, so $\pi x^2 = \pi \sqrt{y/k}$.

$$V = \int_0^y \pi \sqrt{y/k}\, dy = \frac{2\pi}{3\sqrt{k}} y^{3/2}.$$

(b) Torricelli's law: $\dfrac{dV}{dt} = -c\sqrt{y}$. Now

$$\frac{dV}{dt} = \frac{dV}{dy} \cdot \frac{dy}{dt}, \quad \text{so} \quad -c\sqrt{y} = \left(\pi\sqrt{y/k}\right)\frac{dy}{dt}.$$

Therefore $\dfrac{dy}{dt} = -\dfrac{c}{\pi}\sqrt{k}$, a constant.

Section 6.3

1. $V = \displaystyle\int_0^2 2\pi x\left(x^2\right) dx = 8\pi.$

4. $V = \displaystyle\int_0^2 2\pi x\left(8 - 2x^2\right) dx = 16\pi.$

7. $V = \displaystyle\int_0^1 2\pi y\left(3 - 3y\right) dy = \pi.$

10. $V = \displaystyle\int_0^3 2\pi x\left(3x - x^2\right) dx = \frac{27\pi}{2}.$

13. $V = \displaystyle\int_0^1 2\pi x\left(x - x^3\right) dx = \frac{4\pi}{15}.$

16. $V = \displaystyle\int_0^2 2\pi x\left(x^3\right) dx = \frac{64\pi}{5}.$

19. $V = \displaystyle\int_{-1}^1 2\pi(2 - x)\left(x^2\right) dx = \frac{8\pi}{3}.$ This is one of many examples in which you do *not* obtain the correct answer by doubling the value of the integral from 0 to 1.

22. $V = \displaystyle\int_0^1 2\pi(2 - y)\left(\sqrt{y} - y\right) dy = \frac{8\pi}{15}.$

25. $V = \displaystyle\int_{-1}^1 2\pi(1 - y)\left(2 - 2y^2\right) dy = \frac{16\pi}{3}.$

28. $V = \displaystyle\int_0^1 2\pi(y + 1)\left(\sqrt{y} - y^2\right) dy = \frac{29\pi}{30}.$

31. The curves cross at the two points with x-coordinates $b \approx 0.824132312$ and $a = -b$. The volume generated by rotation of R around the y-axis is

$$V = \int_a^b 2\pi x\left[(\cos x) - x^2\right] dx = \pi\left[2x\sin x + 2\cos x - \tfrac{1}{2}x^4\right]_a^b \approx 1.060269.$$

34. The curves cross at the two points $(-a, 0)$ and $(a, 0)$ where $a \approx 1.7878717268$. The volume generated by rotation of R around the y-axis is

$$V = \int_0^a 2\pi x \cdot (3\cos x + \cos 4x)\, dx = \frac{\pi}{8}\left[48\cos x + 48x\sin x + \cos 4x + 4x\sin 4x\right]_0^a \approx 12.004897.$$

37. $V = 2\displaystyle\int_0^a 2\pi x \frac{b}{a}\left(a^2 - x^2\right)^{1/2} dx = \frac{4}{3}\pi a^2 b.$

40. **(a)** $V = \int_{-1}^{2} 2\pi(x+2)\left(x+2-x^2\right) dx = \dfrac{45\pi}{2}.$

(b) $V = \int_{-1}^{2} 2\pi(3-x)\left(x+2-x^2\right) dx = \dfrac{45\pi}{2}.$

43. $a^2 + \frac{1}{4}h^2 = b^2$, so $b^2 - a^2 = \frac{1}{4}h^2$. Therefore $V = \frac{4}{3}\pi\left(\frac{1}{4}h^2\right)^{3/2}$. So the answer to part (a) is $V = \frac{1}{6}\pi h^3$.

(b) The answer involves neither the radius b of the sphere nor the radius a of the hole; it depends only upon the length of the hole.

46. Let $f(x) = x\sqrt{x+3}$. Part (a): The volume is $V_1 = \int_{-3}^{0} \pi \left(f(x)\right)^2 dx = \frac{27}{4}\pi.$

Part (b): The volume is $V_2 = \int_{-3}^{0} 4\pi f(x) dx = \frac{576}{35}\pi\sqrt{3}.$

Part (c): $V_3 = \int_{-3}^{0} -4(x+3)\pi f(x) dx = \frac{432}{35}\pi\sqrt{3}.$

Section 6.4

1. $\displaystyle\int_{0}^{1} \sqrt{1+4x^2}\, dx$

4. $\displaystyle\int_{-1}^{1} \sqrt{1+\frac{16}{9}x^{2/3}}\, dx$

7. $\displaystyle\int_{-1}^{2} \sqrt{1+16y^6}\, dy$

10. $\displaystyle\int_{0}^{2} \frac{2}{\sqrt{4-x^2}}\, dx$

13. $\displaystyle\int_{0}^{1} 2\pi\left(x-x^2\right)\sqrt{4x^2-4x+2}\, dx$

16. $\displaystyle\int_{0}^{1} 2\pi\left(x-x^3\right)\sqrt{9x^4-6x^2+2}\, dx$

19. $\displaystyle\int_{1}^{4} 2\pi(x+1)\sqrt{1+\frac{9}{4}x}\, dx$

22. $\dfrac{dx}{dy} = \sqrt{y-1}$, so $L = \int_{1}^{5} \left(1+\left(\sqrt{y-1}\right)^2\right)^{1/2} dy = \frac{2}{3}\left(5\sqrt{5}-1\right).$

25. $\dfrac{dy}{dx} = \dfrac{6x^4-8y}{4x}$. But $8y = \dfrac{2x^6+1}{x^2}$, so $\dfrac{dy}{dx} = x^3 - \frac{1}{4}x^{-3}$. Therefore $1+\left(\dfrac{dy}{dx}\right)^2 = \left(x^3+\frac{1}{4}x^{-3}\right)^2$, so

$$L = \int_{1}^{2} \left(x^3+\frac{1}{4}x^{-3}\right) dx = \frac{123}{32} = 3.84375.$$

28. $2(y-3)\dfrac{dy}{dx} = 12(x+2)^2$, so $\dfrac{dy}{dx} = \dfrac{6(x+2)^2}{y-3}$. Thus

$$1+\left(\dfrac{dy}{dx}\right)^2 = 1+\dfrac{36(x+2)^2}{(y-3)^2} = 1+\dfrac{36(x+2)^4}{4(x+2)^3} = 1+9(x+2) = 9x+19.$$

Therefore

$$L = \int_{-1}^{2} (9x+19)^{1/2}\, dx = \dfrac{2}{27}\left(37\sqrt{37}-10\sqrt{10}\right).$$

31. $1 + \left(\dfrac{dy}{dx}\right)^2 = \left(x^4 + \frac{1}{4}x^{-4}\right)^2$. Therefore $A = \displaystyle\int_1^2 2\pi x \left(x^4 + \frac{1}{4}x^{-4}\right)\, dx = \frac{339}{16}\pi$.

34. $A = \displaystyle\int_1^2 2\pi x\sqrt{1+x}\, dx$. Let $u = 1+x$; then $x = u - 1$ and $dx = du$. As x takes on values from 1 to 2, u takes on values from 2 to 3. So

$$A = \int_2^3 2\pi(u-1)u^{1/2}\, du = \frac{8}{15}\pi\left(6\sqrt{3} - \sqrt{2}\right).$$

40. Take $y = f(x) = \left(r^2 - x^2\right)^{1/2}$, $-r \le x \le r$. Then $\dfrac{dy}{dx} = -\dfrac{x}{\sqrt{r^2 - x^2}}$. So

$$A = \int_{-r}^r 2\pi\sqrt{r^2 - x^2}\,\sqrt{1 + \frac{x^2}{r^2 - x^2}}\, dx = 2\pi\int_{-r}^r \left(r^2\right)^{1/2}\, dx = 4\pi r^2.$$

Section 6.5

1. $y^{-1/2} = 2x\, dx$; $2y^{1/2} = x^2 + C$; $y(x) = \frac{1}{4}\left(x^2 + C\right)^2$.

4. $y^{-3/2}\, dy = x^{3/2}\, dx$; $-2y^{-1/2} = \frac{2}{5}x^{5/2} + K$; $y^{-1/2} = \frac{1}{5}\left(C - x^{5/2}\right)$; $y(x) = 25\left(C - x^{5/2}\right)^{-2}$.

7. $\left(1 + y^{1/2}\right)\, dy = \left(1 + x^{1/2}\right)\, dx$; $y + \frac{2}{3}y^{3/2} = x + \frac{2}{3}x^{3/2} + K$; $3y + 2y^{3/2} = 3x + 2x^{3/2} + C$.

10. $2y + \frac{3}{2}y^{-2} = x + x^{-1} + C$

13. $4y^3\, dy = 1\, dx$; $y^4 = x + C$. $1 = 0 + C$, so $C = 1$. $y^4 = x + 1$.

16. $y\, dy = x\, dx$; $\frac{1}{2}y^2 = \frac{1}{2}x^2 + K$; $y^2 = x^2 + C$. $25 = 9 + C$, so $C = 16$. $y^2 = x^2 + 16$; that is, $y(x) = \sqrt{x^2 + 16}$. (We take the positive sign at the last step because $y(3) > 0$.)

19. $y^{-2}\, dy = (3x^2 - 1)\, dx$; $-(y^{-1}) = x^3 - x + K$; $y(x) = \dfrac{1}{C + x - x^3}$. $1 = \dfrac{1}{C}$: $y(x) = \dfrac{1}{1 + x - x^3}$.

22. If $P(t) = \left(\frac{1}{2}ktP_0^{1/2}\right)^2$, then

$$\frac{dP}{dt} = 2\left(\frac{1}{2}kt + P_0^{1/2}\right)\left(\frac{1}{2}k\right) = k\left(\frac{1}{2}kt + P_0^{1/2}\right) = kP^{1/2} \quad\text{and}\quad P(0) = \left(P_0^{1/2}\right)^2 = P_0.$$

Thus we have verified that the proposed solution satisfies both the differential equation and the associated initial condition.

25. $P(t) = \dfrac{2}{1 - 2kt}$; $4 = P(3) = \dfrac{2}{1 - 6k}$, so $k = \frac{1}{12}$. Now $P(6)$ is undefined, but $P(t) \to +\infty$ as $t \to 6$ from below. So the rabbit population explodes in the second three-month period.

28. In the notation of Section 6.5 of the text, we have $c = 1$, $g = 32$, $a = \pi/144$, and $A(y) = 9\pi$. So Eq. (20) takes the form

$$9\pi\frac{dy}{dt} = -\frac{\pi}{144}\sqrt{64y}, \quad\text{and hence}\quad y^{-1/2}\frac{dy}{dt} = -\frac{1}{162}.$$

It follows that $y^{1/2} = C - (t/324)$. Therefore

$$y(t) = \left(C - \frac{t}{324}\right)^2.$$

Now $9 = y(0) = C^2$, so $C = 3$ (not -3, else y would never be zero). Thus

$$y(t) = \left(3 - \frac{t}{324}\right)^2$$

where the units are feet and seconds (because those are the units that we used for g and other assorted constants and variables in the problem). The tank will be empty when $y = 0$; that is, when $t = 972$ (s). Hence the tank will require approximately 16 min 12 s to empty.

31. Let $V(y)$ denote the volume of water in the tank when the water has depth y. Then $\dfrac{dV}{dt} = -C_0\sqrt{y}$ and $V(y) = \displaystyle\int_0^y \pi u^{3/2}\, du$. Therefore

$$\pi y^{3/2}\frac{dy}{dt} = -C_0 y^{1/2}$$
$$\pi y\, dy = -C_0\, dt$$
$$\frac{1}{2}\pi y^2 = -C_0 t + K_0$$
$$y^2 = -Kt + C.$$

Let $t = 0$ at noon. Then $y(0) = 12$, so $C = 144$. Since $y(1) = 6$, $36 = 144 - K$, so $K = 108$. Therefore $y^2 = 144 - 108t$. When $y = 0$, $t = 4/3$. The tank will be empty at 1:20 P.M.

34. Set up a coordinate system in which the tank has its lowest point at the origin and its vertical diameter lying on the y-axis, so that the equation of the cross section of the tank in the xy-plane will be $x^2 + (y - 4)^2 = 16$. If the liquid in the tank has depth y, then the radius at its surface is $x = (8y - y^2)^{1/2}$, so in Eq. (20) we take $A(y) = \pi x^2\,(8y - y^2)$, $c = 1$, $g = 32$, and $a = \pi/144$ to obtain

$$A(y) = \pi x^2 = \pi\,(8y - y^2)\,.$$

Thus

$$\pi\,(8y - y^2)\,\frac{dy}{dt} = -\frac{\pi}{144}\sqrt{64y}\,:$$
$$\left(8y^{1/2} - y^{3/2}\right)dy = -\frac{1}{18}\,dt;$$
$$\frac{16}{3}y^{3/2} - \frac{2}{5}y^{5/2} = -\frac{t}{18} + C.$$

When $t = 0$, $y = 8$. It follows that $C = \frac{512}{15}\sqrt{2}$. Thus

$$\frac{16}{3}y^{3/2} - \frac{2}{5}y^{5/2} = -\frac{t}{18} + \frac{512}{15}\sqrt{2}.$$

The tank is empty when $t = 0$. At that time we have $t = \frac{3072}{5}\sqrt{2} \approx 869$ seconds—about 14 minutes 29 seconds.

Section 6.6

1. $W = \displaystyle\int_{-2}^{1} 10\, dx = 30.$

4. $W = \displaystyle\int_0^4 -3\sqrt{x}\, dx = -16.$

7. $W = \displaystyle\int_0^1 30x\, dx = 15 \text{ (ft·lb)}.$

10. $W = \displaystyle\int_0^{10} (62.4)(y)(25\pi)\, dy = 78000\pi \approx 245{,}044.23 \text{ (ft·lb)}.$

13. Take $y = -10$ as ground level, so that the radius x of a thin horizontal slab of water at height y above ground level satisfies the equation $x^2 = 5y$. Therefore

$$W = \int_0^5 (y + 10)(50)(5\pi y)\, dy = \frac{125{,}000}{3}\pi \approx 130{,}900 \text{ (ft·lb)}.$$

16. Set up the following coordinate system: The x-axis and y-axis cross at the center of one end of the tank, so that the equation of the circular (vertical) cross section of the tank is $x^2 + y^2 = 9$. Then the gasoline must be lifted to the level $y = 10$. A horizontal cross section of the tank at the level y is a rectangle of length 10 and width $2x$ where x and y satisfy the equation $x^2 + y^2 = 9$ of the end of the tank. Thus we find that $x = (9 - y^2)^{1/2}$. So

(a) The amount of work required to pump all of the gasoline into the cars is:

$$W = \int_{-3}^3 \left(2\sqrt{9 - y^2}\right)(10)(10 - y)(45)\, dy$$

$$= 900 \int_{-3}^3 \left(10\sqrt{9 - y^2} - y\sqrt{9 - y^2}\right) dy$$

$$= 9000 \cdot \frac{1}{2} \cdot \pi \cdot 9 + 900 \left[\frac{1}{3}(9 - y^2)^{3/2}\right]_{-3}^3$$

$$= 40500\pi \approx 127{,}234.5 \text{ (ft·lb)}.$$

(b) Assume that the tank has a 1.341 hp motor. If it were to operate at 100% efficiency, it would pump all of the gasoline in $40500\pi/((3300)(1.341)) \approx 2.875161$ min, using 1 kW for 2.875161 minutes. This would amount to 0.047919356 kWh, costing 0.345¢. Assuming that the pump in the tank is only 30% efficient, the actual cost would be about 1.15¢.

19. Take $y = 0$ at ground level. At time t, $0 \le t \le 50$, the bucket is at height $y = 2t$ and its weight is $F = 100 - \frac{1}{2}t$. So $t = \frac{1}{2}y$, and therefore the work is given by

$$W = \int_0^{100} \left(100 - \frac{1}{4}y\right) dy = 8750 \text{ (ft·lb)}.$$

22. $W = \displaystyle\int_{x_1}^{x_2} Ap(Ax)\, dx.$ Let $V = Ax$; then $dV = A\, dx$. Therefore

$$W = \int_{V_1}^{V_2} pV\, dV.$$

25. $W = \displaystyle\int_0^1 60\pi(1 - y)\sqrt{y}\, dy = 16\pi$ (ft·lb).

28. There are a number of ways to work this problem. Here is a way to check your answer by using elementary physics and no calculus. Stage 1: Imagine the chain hanging in the shape of an "L" with 40 feet vertical and 10 feet horizontal, the latter on the floor of the monkey's cage in a neat heap. Stage 2: Cut off this ten-foot length of chain and move it to its final position—hanging from the top of the cage. It weighs 5 pounds and has moved an average distance of 35 feet, so the work to lift this segment of the chain is $S_1 = (5)(35) = 175$ ft·lb. Stage 3: Return to the dangling chain. Cut off its bottom 15 feet and move it to its final position—hanging from the 10-foot segment now hanging from the top of the cage. The top end of this segment is lifted $30 - 15 = 15$ feet and it weighs 7.5 pounds, so the work to lift the second segment is $S_2 = (15)(7.5) = 112.5$ ft·lb. The remaining 25 feet of the chain doesn't move at all. Finally, the monkey lifts her own 20 pounds a distance of 40 feet, so the work involved here is $S_3 = 800$ ft·lb. So the total work in the process is

$W = W_1 + W_2 + W_3 = 1087.5$ ft·lb. Of course, to gain the maximum benefit from this problem, you should work it using techniques of calculus.

31. Set up a coordinate system with the y-axis vertical and the x-axis coinciding with the bottom of one end of the trough. A horizontal section of the trough at y is $2 - y$ feet below the water surface, so the total force on the end of the trough is given by

$$F = \int_0^2 (2)(2 - y)(\rho)\, dy = 6\rho = 249.6 \text{ (pounds)}.$$

34. Describe the end of the tank by the inequality $x^2 + y^2 \le 16$, so that a horizontal section at level y has width $2x = 2\left(16 - y^2\right)^{1/2}$. Then the total force on the end of the tank is

$$F = \int_{-4}^{4} \rho(2)(4 - y)\left(16 - y^2\right)^{1/2} dy$$

$$= 2\rho \int_{-4}^{4} 4\left(16 - y^2\right)^{1/2} dy - 2\rho \int_{-4}^{4} y\left(16 - y^2\right)^{1/2} dy$$

$$= (16\rho)(4\pi) = 64\rho\pi = 3200\pi \approx 10053.1 \text{ (pounds)}.$$

37. Place the x-axis along the water surface. Then a horizontal section of the triangle at level y has width $2x = \frac{8}{5}(y + 15)$, and hence the total force on the gate is

$$F = \int_{-15}^{-10} -\frac{8}{5}\rho y(y + 15)\, dy.$$

Chapter 6 Miscellaneous

1. Note that $t^2 - t - 2 = (t + 1)(t - 2)$ is negative on the interval $(-1, 2)$. Therefore the net distance traveled is

$$\int_0^3 (t^2 - t - 2)\, dt = -\frac{3}{2},$$

whereas the total distance traveled is

$$-\int_0^2 (t^2 - t - 2)\, dt + \int_2^3 (t^2 - t - 2)\, dt = \frac{31}{6}.$$

4. $V = \int_0^1 x^3\, dx = \frac{1}{4}.$

7. $V = \int_0^1 \pi\left(x^2 - x^4\right) dx = \pi\left(\frac{1}{3} - \frac{1}{5}\right) = \frac{2}{15}\pi.$

10. $V = \int_0^1 \left(2x - x^2 - x^3\right)^2 dx = \frac{22}{105}.$

13. $M = \int_0^{20} \frac{\pi}{16}(8.5)\, dx = (10.625)\pi \approx 33.379$ (grams).

16. Because $(a - h, r)$ lies on the ellipse, $\left(\dfrac{a - h}{a}\right)^2 + \left(\dfrac{r}{b}\right)^2 = 1$. Therefore $r^2 = \dfrac{2ah - h^2}{a^2} b^2$. And so

$$V = \int_{a-h}^{a} \pi y^2\, dx = \int_{a-h}^{a} \pi b^2 \left(1 - \frac{x^2}{a^2}\right) dx = \pi \frac{b^2 h^2}{3a^2}(3a - h).$$

But $r^2 = \dfrac{b^2}{a^2} h(2a - h)$, so $\dfrac{b^2}{a^2} h = \dfrac{r^2}{2a - h}$. Therefore

$$V = \tfrac{1}{3}\pi r^2 h\, \frac{3a - h}{2a - h}.$$

19. $V = \displaystyle\int_1^t \pi\,(f(x))^2\, dx = \frac{\pi}{6}\left[(1 + 3t)^2 - 16\right]$. Thus

$$\pi\,(f(x))^2 = \frac{\pi}{6}\left[(2)(1 + 3x)(3)\right] = \pi(1 + 3x).$$

Therefore $f(x) = \sqrt{1 + 3x}$.

22. If $-1 \le x \le 2$, then a thin vertical strip of the region above x is rotated in a circle of radius $x + 2$. Therefore the volume generated is

$$V = \int_{-1}^{2} 2\pi(x + 2)(x + 2 - x^2)\, dx = \tfrac{45}{2}\pi.$$

25. $\dfrac{dx}{dy} = \tfrac{3}{8}\left(\tfrac{4}{3}y^{1/3} - \tfrac{4}{3}y^{-1/3}\right)$. So—after some algebra—

$$\left(1 + \left(\frac{dx}{dy}\right)^2\right)^{1/2} = \tfrac{1}{2}\left(y^{1/3} + y^{-1/3}\right).$$

Therefore

$$L = \tfrac{1}{2}\int_1^8 \left(y^{1/3} + y^{-1/3}\right)\, dy = \tfrac{63}{8}.$$

28. $\dfrac{dy}{dx} = -\dfrac{x}{\sqrt{r^2 - x^2}}$, so $1 + \left(\dfrac{dy}{dx}\right)^2 = \dfrac{r^2}{r^2 - x^2}$. Therefore

$$A = \int_a^b 2\pi\sqrt{r^2 - x^2}\,\frac{r}{\sqrt{(r^2 - x^2)}}\, dx = 2\pi r h.$$

34. $(y + 1)^{-1/2}\, dy = 1\, dx$;

$2(y + 1)^{1/2} = x + C$;

$y(x) = -1 + \tfrac{1}{4}(x + C)^2$.

37. $y^2\, dy = \dfrac{1}{x^2}\, dx$;

$\tfrac{1}{3}y^3 = -\dfrac{1}{x} + K$;

$y^3 = C - \dfrac{3}{x}$;

$y(x) = \left(C - \dfrac{3}{x}\right)^{1/3}$.

40. $y^{-1/2}\, dy = \sin x\, dx$;

$2y^{1/2} = C - \cos x$;

$y(x) = \tfrac{1}{4}(C - \cos x)^2$. $\qquad 4 = y(0) = \tfrac{1}{4}(C - 1)^2$: $\; C = 5$ or $C = -3$.

Two solutions: $y(x) = \tfrac{1}{4}(5 - \cos x)^2$, $\; y(x) = \tfrac{1}{4}(3 + \cos x)^2$.

43. Denote by K the spring constant. Then

$$\int_2^5 K(x - L)\, dx = 5 \int_2^3 K(x - L)\, dx.$$

After applying the fundamental theorem of calculus, we find that

$$(5 - L)^2 - (2 - L)^2 = 5(3 - L)^2 - 5(2 - L)^2.$$

Solve for L: $L = 1$, and therefore the natural length of the spring is 1 foot.

46. Set up a coordinate system with the axis of the cone lying on the y-axis and with a diameter of the cone lying on the x-axis. Now a horizontal slice of the cone at height y has radius given by $x = \frac{1}{2}(1 - y)$; the units here are in feet. Therefore the work done in building the anthill is

$$W = \int_0^1 \tfrac{1}{4}(150y)\pi(1 - y)^2\, dy = \tfrac{25}{8}\pi \approx 9.82 \quad \text{(ft·lb)}.$$

49. If the coordinate system is chosen with the origin at the midpoint of the bottom of the dam and with the x-axis horizontal, then the equation of the slanted edge of the dam is $y = 2x - 200$ (with units in feet). Therefore the width of the dam at level y is $2x = y + 200$. Thus the total force on the dam is

$$F = \int_0^{100} \rho(100 - y)(y + 200)\, dy.$$

Chapter 7: Exponential and Logarithmic Functions
Section 7.1

1. **(a)** $2^3 \cdot 2^4 = 2^{3+4} = 2^7 = 128$.   **(b)** $3^2 \cdot 3^3 = 3^5 = 243$.
    (c) $\left(2^2\right)^3 = 64$.   **(d)** $2^{2^3} = 2^8 = 256$.
    (e) $3^5 \cdot 3^{-5} = 3^{5-5} = 3^0 = 1$.

4. **(a)** $\log_7 7^2 = 2\log_7 7 = 2$.   **(b)** $\log_{10} 1000 = \log_{10} 10^3 = 3\log_{10} 10 = 3$.
    (c) $\log_{12} 12^2 = 2\log_{12} 12 = 2$.

7. **(a)** $\ln\frac{8}{27} = (\ln 2^3) - (\ln 3^3) = 3\ln 2 - 3\ln 3$.   **(b)** $(2\ln 2) + (\ln 3) - (2\ln 5)$.

10. If $x = \log_{0.5} 16$, this means that $(0.5)^x = 16$. Take the reciprocal of each side in the last equation to obtain
$$2^x = \frac{1}{16} = \frac{1}{2^4} = 2^{-4},$$
then take the base 2 logarithm of the first and last terms above to conclude that $x = -4$. Finally verify this answer by substitution (because the method is predicated on the assumption that x exists—that is, that the logarithm to the base 1/2 of 16 exists).

13. $2^x < e^x < 3^x$ if $x > 0$, whereas $3^x < e^x < 2^x$ if $x < 0$.

16. $(0.2)^x < (0.4)^x < (0.6)^x$ if $x < 0$, whereas $(0.6)^x < (0.4)^x < (0.2)^x$ if $x > 0$.

19. $10^{-x} = 10^2$: $x = -2$.

22. $\log_x 16 = 2$ implies that $x^2 = 16$, so that $x = 4$ (only positive numbers other than 1 may be used as bases for logarithms).

25. Apply the natural logarithm after division of both sides by 3 to obtain $\ln(e^x) = \ln 1$, so that $x = 0$.

28. $\dfrac{dy}{dx} = x^3 e^x + 3x^2 e^x$.

31. $\dfrac{dy}{dx} = \dfrac{x-2}{x^3} e^x$.

34. $\dfrac{dy}{dx} = x + 2x \ln x$.

37. $\dfrac{dy}{dx} = \dfrac{1-x}{e^x}$.

40. $y = \frac{1}{3}\ln x$, so $\dfrac{dy}{dx} = \dfrac{1}{3x}$.

43. $\dfrac{dy}{dx} = e^{3x}(3\cos 4x - 4\sin 4x)$.

46. $y = 2x + 1$.

49. $f^{(n)}(x) = 2^n e^{2x}$.

52. $f'(x) = 0$ when $x = 5$, $f'(x) > 0$ on $(0,5)$, and $f'(x) < 0$ if $x > 5$. Therefore the global maximum value of f is $f(5) = 2e^{-5/2}$.

55. First, $f'(x) = \dfrac{6\cos x - \sin x}{6e^{x/6}}$, so the first local maximum point for $x > 0$ occurs when $x = \arctan 6$ and the first local minimum point when $x = \pi + \arctan 6$. The corresponding y-coordinates are, respectively,
$$\frac{6}{e^{(\arctan 6)/6}\sqrt{37}} \quad \text{and} \quad -\frac{6}{e^{(\pi+\arctan 6)/6}\sqrt{37}}.$$

58. Part (a): $P(t) = 4 \cdot 2^{t/3} = 12$ when $t = (\ln 27)(\ln 2) \approx 4.75489$. So it requires about 4.75 months for the population to triple. Part (b): At the end of one year the population numbers $P(12) = 64$ rabbits and is then growing at the rate of $P'(12) = \frac{64}{3}\ln 2 \approx 14.7871$ rabbits per month.

Section 7.2

1. $f'(x) = \dfrac{1}{3x-1} D_x(3x-1) = \dfrac{3}{3x-1}$

4. $f(x) = 2\ln(1+x)$, so $f'(x) = \dfrac{2}{1+x}$.

7. $f'(x) = -\dfrac{1}{x}\sin(\ln x)$

10. $f'(x) = \dfrac{1}{\ln x} D_x(\ln x) = \dfrac{1}{x\ln x}$

16. $f'(x) = [\cos(\ln 2x)] D_x(\ln 2x) = \dfrac{1}{x}\cos(\ln 2x)$

19. $f(x) = 3\ln(2x+1) + 4\ln(x^2-4)$, so $f'(x) = \dfrac{6}{2x+1} + \dfrac{8x}{x^2-4}$.

22. $f(x) = \frac{1}{2}\ln(4x-7) - 3\ln(3x-2)$, so $f'(x) = \dfrac{2}{4x-7} - \dfrac{9}{3x-2} = \dfrac{59-30x}{(4x-7)(3x-2)}$.

25. *Suggestion:* Write $g(t) = 2\ln t - \ln(t^2+1)$.

28. $f(x) = \ln(\sin x) - \ln(\cos x)$, so $f'(x) = \dfrac{\cos x}{\sin x} + \dfrac{\sin x}{\cos x} = \cot x + \tan x$.

34. $\displaystyle\int \dfrac{dx}{3x+5} = \frac{1}{3}\ln|3x+5| + C$.

40. $\displaystyle\int \dfrac{1}{x\ln x}\,dx = \ln|\ln x| + C$.

46. $\displaystyle\int \dfrac{\ln(x^3)}{x}\,dx = 3\int \dfrac{\ln x}{x}\,dx = \frac{3}{2}(\ln x)^2 + C$.

49. The numerator is $\frac{1}{3}$ the derivative of the denominator.

52. $\displaystyle\lim_{x\to\infty} \dfrac{\ln x^3}{x^2} = \lim_{x\to\infty} \dfrac{3\ln x}{x^2} = 0$.

55. *Suggestion:* Make the substitution $x = \dfrac{1}{u^2}$.

58. $f^{(1)}(x) = x^{-1}$, $f^{(2)}(x) = (-1)x^{-2}$, $f^{(3)}(x) = (-2)(-1)x^{-3}$, $f^{(4)}(x) = (-3)(-2)(-1)x^{-4}$, ..., and, in general, $f^{(n)}(x) = (-1)^{n-1}(n-1)!\,x^{-n}$ for $n \geq 1$. Proof: By induction on n.

64. Two solutions, 1.615 and 5.629.

70. The curves cross near $x = 2.1956108$ and $x = 9.7747251$. The area between them is approximately 86.1489.

73. Let $y = 1/x$; $x = 1/y$. Then $\displaystyle\lim_{x\to 0^+} x^k \ln x = \lim_{y\to\infty} \dfrac{\ln(1/y)}{y^k} = -\lim_{y\to\infty} \dfrac{\ln y}{y^k} = 0$.

76. $\dfrac{dy}{dx} = \dfrac{2+\ln x}{2x^{1/2}}$ and $\dfrac{d^2y}{dx^2} = -\dfrac{\ln x}{4x^{3/2}}$. Now $\dfrac{dy}{dx} = 0$ when $x = 1/(e^2)$ (there, $y = -2/e$). As $x \to 0^+$, $y \to 0$ while $\dfrac{dy}{dx} \to -\infty$. The graph is increasing for $x > 1/(e^2)$ and decreasing for $0 < x < 1/(e^2)$. It is concave upward on $(0,1)$, concave downward if $x > 1$. The point $(1,0)$ is an inflection point and the only intercept. The point where there is a horizontal tangent is in fact a global minimum. Note that the point $(0,0)$ is not on the graph. It is also helpful to note that $dy/dx = 1$ when $x = 1$.

79. $W = \displaystyle\int_{V_1}^{V_2} p(V)\,dV = \int_{V_1}^{V_2} \dfrac{nRT}{V}\,dV = nRT\,(\ln(V_2) - \ln(V_1)) = nRT\ln\left(\dfrac{V_1}{V_2}\right)$.

Section 7.3

1. $f'(x) = e^{2x} D_x(2x) = 2e^{2x}$

4. $f'(x) = -3x^2 e^{4-x^3}$

10. $g'(t) = \frac{1}{2}(e^t - e^{-t})^{-1/2}(e^t + e^{-t})$

16. $f'(x) = e^x \cos 2x - 2e^x \sin 2x$

22. $g'(t) = \dfrac{2e^t}{(1 - e^t)^2}$

28. $f'(x) = (e^{2x} + e^{-2x})^{-1/2}(e^{2x} - e^{-2x})$

34. $\dfrac{dy}{dx} = \dfrac{1}{(1+y)e^y} = \dfrac{y}{x(1+y)}$

40. $\displaystyle \int \sqrt{x}\, e^{2x\sqrt{x}} = \frac{1}{3}e^{2x\sqrt{x}} + C$

46. $\displaystyle \int e^{2x+3}\, dx = \frac{1}{2}e^{2x+3} + C$

49. Suggestion: Let $u = \sqrt{x}$.

52. Because $e^{a+b} = e^a e^b$, we have $\displaystyle \int e^{x+e^x}\, dx = \int e^x\, e^{e^x}\, dx = e^{e^x} + C$. Note that e^{e^x} means $e^{(e^x)}$.

58. $\displaystyle \lim_{h \to 0}(1 + 2h)^{1/h} = e^2$

64. First, $dy/dx = 3x^2 e^{-x} - x^3 e^{-x} = x^2(3-x)e^{-x}$
and $d^2y/dx^2 = x(x^2 - 6x + 6)e^{-x}$. Note that $y = 0$
when $x = 0$ and that this is the only intercept. Next,
$dy/dx = 0$ when $x = 0$ and when $x = 3$, so there are
horizontal tangents at $(0,0)$ and at $\left(3, \dfrac{27}{e^3}\right)$—near
the point $(3, 1.344)$. The second derivative vanishes
when $x = 0$ and when $x^2 - 6x + 6 = 0$; the latter
equation has the two solutions $x = 3 \pm \sqrt{3}$. Conse-
quently there may be inflection points at $(0,0)$, near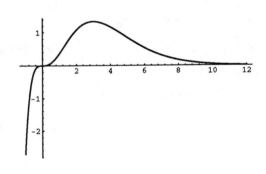
$(1.268, 0.574)$, and near $(4.732, 0.933)$. The graph is
increasing if $x < 3$ and decreasing if $x > 3$. It is concave downward if $x < 0$ and if $|x - 3| < \sqrt{3}$,
concave upward if **both** $|x - 3| > \sqrt{3}$ and $x > 0$. Because $x^3 e^{-x} \to 0$ as $x \to +\infty$, the x-axis is a
horizontal asymptote. The graph is shown above.

67. $V = \displaystyle \int_0^1 \pi e^{2x}\, dx = \frac{\pi}{2}(e^2 - 1) \approx 10.0359$.

70. $A = \displaystyle \int_0^1 (2\pi)\left[\frac{1}{2}(e^x + e^{-x})\right]\left[\frac{1}{2}(e^x + e^{-x})\right]\, dx = \frac{\pi}{2}\int_0^1 (e^{2x} + 2 + e^{-2x})\, dx$
$\quad = \dfrac{\pi}{4}(e^2 + 4 - e^{-2}) \approx 8.83865$.

76. Three solutions: 0.291, 0.989, and 1.493.

82. $e - 1 = \displaystyle \int_0^1 e^x\, dx \approx \frac{1}{6}(e^0 + 4\sqrt{e} + e)$;

$$6e - 6 \approx 1 + 4\sqrt{e} + e;$$
$$5e - 7 \approx 4\sqrt{e};$$
$$25e^2 - 70e + 49 \approx 16e;$$
$$25e^2 - 86e + 49 \approx 0;$$
$$e \approx \tfrac{1}{50}\left(86 + \sqrt{2496}\right);$$

therefore $e \approx 2.7192$.

85. Given $y''(x) + y'(x) - 2y(x) = 0$, we write the associated equation $m^2 + m - 2 = 0$, which has solutions $m_1 = -2$ and $m_2 = 1$. Therefore the function $y(x) = Ae^{-2x} + Be^x$ satisfies the given differential equation for any choice of the constants A and B. The additional information given in the problem is this:

$$5 = y(0) = A + B,$$
$$2 = y'(0) = -2A + B.$$

It follows that $A = 1$ and $B = 4$, and therefore a solution of the given differential equation satisfying the two additional side conditions is $y(x) = e^{-2x} + 4e^x$.

Section 7.4

1. $f'(x) = 10^x \ln 10$

4. $f(x) = (\log_{10} e)(\ln \cos x) = \dfrac{1}{\ln 10}\ln(\cos x)$, so $f'(x) = \left(\dfrac{1}{\ln 10}\right)\left(\dfrac{-\sin x}{\cos x}\right)$.

10. $f'(x) = 7^{(8^x)}(\ln 7)(8^x \ln 8)$

16. $f(x) = (\log_2 e)(\log_e x) = \dfrac{\ln x}{\ln 2}$, so $f'(x) = \dfrac{1}{x \ln 2}$.

22. $f'(x) = \pi^x \ln \pi + \pi x^{\pi - 1}$

28. The substitution $u = 1/x$ yields

$$\int \frac{10^{1/x}}{x^2}\, dx = \int -10^u\, du = -\frac{10^u}{\ln 10} + C = -\frac{10^{1/x}}{\ln 10} + C.$$

31. Write $\log_2 x$ as $(\ln x)/(\ln 2)$. Then, if you wish, use the substitution $u = \ln x$.

34. $\ln y = \tfrac{1}{3}\ln\left(3 - x^2\right) - \tfrac{1}{4}\ln\left(x^4 + 1\right)$, so

$$\frac{1}{y}\cdot\frac{dy}{dx} = \frac{-2x}{3(3 - x^2)} - \frac{x^3}{x^4 + 1};$$

$$\frac{dy}{dx} = -\frac{(3 - x^2)^{1/2}}{(x^4 + 1)^{1/4}}\left(\frac{x}{(3 - x^2)} + \frac{x^3}{x^4 + 1}\right).$$

37. $\ln y = (\ln x)^2$. So $\dfrac{dy}{dx} = (y)\dfrac{2\ln x}{x} = \left(x^{\ln x}\right)\dfrac{2\ln x}{x}$.

40. $\dfrac{dy}{dx} = (y)\left(\dfrac{1}{2(x + 1)} + \dfrac{1}{3(x + 2)} + \dfrac{1}{4(x + 3)}\right)$

$$= \sqrt{x + 1}\sqrt[3]{x + 2}\sqrt[4]{x + 3}\left(\frac{1}{2(x + 1)} + \frac{1}{3(x + 2)} + \frac{1}{4(x + 3)}\right).$$

46. $\ln y = x \ln \left(1 + \dfrac{1}{x}\right) = x \ln(x+1) - x \ln x.$

$$\frac{dy}{dx} = \left(1 + \frac{1}{x}\right)^x \left(\ln(x+1) + \frac{x}{x+1} - \ln x - 1\right).$$

52. The curves cross at $x = 0$ and at $x = a \approx 1.578620636$, and the area between them is

$$\int_0^a \left(2^{-x} - (x-1)^2\right) \, dx \approx 0.561770.$$

55. $f'(x) = \dfrac{\cos x - (x \ln x)\sin x}{x} \cdot x^{\cos x}.$

58. If $y = u^v$ where all are functions of x, then $\ln y = v \ln u$. With $u'(x)$ denoted simply by u', etc., we now have

$$\frac{1}{y} y' = v' \ln u + \frac{vu'}{u}.$$

Thus $y' = u^v \, v' \ln u + \dfrac{u^v vu'}{u} = vu^{v-1}u' + u^v(\ln u)v'.$

(i) If u is constant, this implies that $\dfrac{dy}{dx} = u^{v(x)}(\ln u)v'(x).$

(ii) If v is constant, this implies that $\dfrac{dy}{dx} = v\left(u(x)\right)^{v-1} u'(x).$

61. $\ln \left(\dfrac{x^x}{e^x}\right) = x \ln \left(\dfrac{x}{e}\right).$ The latter clearly approaches $+\infty$ as x does, and this is enough to establish the desired result.

Section 7.5

1. The principal at time t is $P(t) = 1000e^{(0.08)t}$. So $P'(t) = 80e^{(0.08)t}$.

Answers: $P'(5) = \$119.35$ and $P'(20) = \$396.24$.

4. Suppose that the skull was formed at time $t = 0$ (in years). Then the amount of ^{14}C it contains at time t will be

$$Q(t) = Q_0 e^{-kt}$$

where Q_0 is the initial amount and $k = \dfrac{\ln 2}{5700} \approx 0.0001216$. We find the value $t = T$ corresponding to "now" by solving

$$Q(T) = \tfrac{1}{6}Q_0 : \quad \tfrac{1}{6}Q_0 = Q_0 e^{-kT}, \text{ so } 6 = e^{kT};$$

therefore $T = \dfrac{1}{k} \ln 6 = 5700 \dfrac{\ln 6}{\ln 2} \approx 14734$, the approximate age of the skull in years.

7. $1 + r = \left(1 + \dfrac{0.09}{4}\right)^4$ for quarterly compounding, and in this case we find that (a) $r \approx 9.308\%$. With the aid of similar formulas, we obtain the other four answers.

10. If $C(t)$ is the concentration of the drug at time t (hours), then

$$C(t) = C_0 e^{-kt}$$

where $k = \tfrac{1}{5} \ln 2$. We require C_0 so large that $C(1) = (45)(50) = 2250$. Thus $C_0 e^{-k} \geq 2250$; that is, $C_0 \geq 2250e^k \approx 2584.57$ (mg).

13. Let Q denote the amount of radioactive cobalt remaining at time t (in years), with the occurrence of the accident set at time $t = 0$. Then

$$Q = Q_0 e^{-(t \ln 2)/(5.27)}.$$

If T is the number of years until the level of radioactivity has dropped to a hundredth of its initial value, then

$$\frac{1}{100} = e^{-(T \ln 2)/(5.27)},$$

and it follows that $T = (5.27) \dfrac{\ln 100}{\ln 2} \approx 35$ (years).

16. Let $T = T(t)$ denote the temperature at time t; by Newton's law of cooling, we have

$$\frac{dT}{dt} = k(T - A).$$

Here, $A = 0$, $T(0) = 25$, and $T(20) = 15$. Also $\dfrac{dT}{dt} = kT$, so

$$T(t) = T_0 e^{kt} = 25 e^{kt}.$$

Next, $15 = T(20) = 25 e^{20k}$, so $k = \frac{1}{20} \ln \left(\frac{3}{5} \right)$. Now $T(t) = 5$ when $5 = 25 e^{kt}$; that is, when

$$t = -\frac{20 \ln 5}{\ln \left(\frac{3}{5} \right)} \approx 63.01.$$

Answer: The buttermilk will be at 5°C about one hour and three minutes after putting it on the porch.

19. We begin with the equation $p(x) = (29.92)e^{-x/5}$.

  (a) $p\left(\frac{10000}{5280}\right) \approx 20.486$ (inches); $p\left(\frac{30000}{5280}\right) \approx 9.604$ (inches).

  (b) If x is the altitude in question, then we must solve

$$15 = (29.92)e^{-x/5};$$
$$x = 5 \ln \left(\frac{29.92}{15} \right) \approx 3.4524 \text{ (miles)},$$

approximately 18230 feet.

22. The projected revenues are $r(t) = (1.85)e^{(0.03)t}$ and the projected budget is $b(t) = 2e^{kt}$ for some constant k (values of both functions are in billions of dollars; remember that in the U.S., a billion is a *thousand* million). In order that $r(7) = b(7)$, we solve to find $k \approx 0.0188626$, so the annual budget increase should be approximately 1.886%.

Section 7.6

1. Given: $\dfrac{dy}{dx} = y + 1$; $y(0) = 1$.

$$\frac{1}{y+2} \frac{dy}{dx} = 1;$$
$$\ln(y+1) = x + C;$$
$$y + 1 = Ke^x.$$

When $x = 0$, $y = 1$. Therefore $2 = (K)(1) = K$, so the solution is $y(x) = -1 + 2e^x$.

4. $\dfrac{dy}{dx} = \dfrac{1}{4} - \dfrac{y}{16} = -\dfrac{1}{16}(y-4); \quad y(0) = 20.$

$$\frac{1}{y-4}\frac{dy}{dx} = -\frac{1}{16};$$
$$\ln(y-4) = C_1 - \frac{x}{16};$$
$$y = 4 + Ce^{-x/16}.$$

We are given $20 = y(0) = 4 + C$, so $C = 16$. Answer: $y(x) = 4 + 16e^{-x/16}$.

10. $v(t) = 10 - 20e^{5t}$

16. Let $Q(t)$ be the number of pounds of salt in the tank at time t (in seconds). Now $Q(0) = 50$, and

$$\frac{dQ}{dt} = -\frac{5}{1000}Q(t) = -\frac{1}{200}Q(t).$$

Therefore

$$Q(t) = 50e^{-t/200}.$$

Next, $Q(t) = 10$ when $e^{-t/200} = \frac{1}{5}$, so that $t = 200\ln 5$. So 10 pounds of salt will remain in the tank after approximately 5 minutes and 22 seconds.

19. In the notation of Problem 18, we have $v = v_0 e^{-kt}$ where we are given $v_0 = 40$. Here, also, $v(10) = 20$, so that $20 = 40e^{-10k}$. It follows that $k = (0.1)\ln 2$, so by Part (b) of Problem 18 the total distance traveled will be $\dfrac{1}{k}v_0 = \dfrac{(40)(10)}{\ln 2}$, approximately 577 feet.

22. $\dfrac{dx}{dt} + ax = be^{ct}, \quad x(0) = x_0, \quad a + c \neq 0.$

$$e^{at}\frac{dx}{dt} + axe^{at} = be^{(a+c)t}.$$
$$D_t\left(e^{at}x(t)\right) = be^{(a+c)t}.$$
$$e^{at}x(t) = C + \frac{b}{a+c}e^{(a+c)t} \qquad (a+c \neq 0).$$

So $x(t) = Ce^{-at} + \dfrac{b}{a+c}e^{ct}$, and $x_0 = x(0) = C + \dfrac{b}{a+c}$, and hence $C = x_0 - \dfrac{b}{a+c}$.

Therefore $x(t) = \left(x_0 - \dfrac{b}{a+c}\right)e^{-at} + \dfrac{b}{a+c}e^{ct} = x_0 e^{-at} + \dfrac{b}{a+c}\left(e^{ct} - e^{-at}\right).$

25. $\dfrac{dx}{dt} = k(100000 - x(t)):$

$$\frac{-dx}{100000 - x} = -k\,dt;$$
$$\ln(100000 - x) = C_1 - kt;$$
$$100000 - x = Ce^{-kt};$$
$$x(t) = 100000 - Ce^{-kt}.$$

On March 1, $t = 0$ and $x = 20000$. On March 15, $t = 14$ and $x = 60000$.

$$20000 = 100000 - C, \text{ so } C = 80000$$
$$x(t) = 10000\left(10 - 8e^{-kt}\right).$$
$$60000 = x(14) = 10000(10 - 8e^{-14k}). \text{ so } 6 = 10 - 8e^{-14k}. \text{ Solve for } k: k = \tfrac{1}{14}\ln 2.$$

(a) $x(t) = 10000(10 - 8e^{-kt})$ where $k = \frac{1}{14}\ln 2$.

(b) $x(T) = 80000$: Solve $10 - 8e^{-kT} = 2$ for T: $T = \dfrac{1}{k}\ln 4 = 28$. So 80,000 people will be infected on March 29.

(c) $\lim\limits_{t\to\infty} N(t) = 100{,}000$: Eventually everybody gets the flu.

Chapter 7 Miscellaneous

1. $f(x) = \ln 2\sqrt{x} = \ln 2 + \frac{1}{2}\ln x$, so $f'(x) = \dfrac{1}{2x}$.

4. $f'(x) = \frac{1}{2}\left(x^{-1/2}\right)10^{\left(x^{1/2}\right)}\ln 10$

7. $f'(x) = 3x^2 e^{-1/x^2} + x^3 e^{-1/x^2}\left(2x^{-3}\right) = \left(2 + 3x^2\right)e^{-1/x^2}$.

10. $f'(x) = (\exp(10^x))(10^x)(\ln 10)$

13. $f'(x) = \dfrac{(x-1) - (x+1)}{(x-1)^2}\exp\left(\dfrac{x+1}{x-1}\right) = -\dfrac{2}{(x-1)^2}\exp\left(\dfrac{x+1}{x-1}\right)$

16. $f'(x) = \dfrac{1}{x}\cos(\ln x)$

19. $f'(x) = \dfrac{3^x \cos x + (3^x \ln 3)\sin x}{3^x \sin x} = \dfrac{\cos x + (\ln 3)\sin x}{\sin x} = \cot x + \ln 3$.

22. If $y = x^{\sin x}$ then $\ln y = (\sin x)(\ln x)$.

So $\dfrac{dy}{dx} = (y)\left(\dfrac{1}{x}\sin x + (\cos x)(\ln x)\right) = (x^{\sin x})\left(\dfrac{1}{x}\sin x + (\cos x)(\ln x)\right)$.

28. $\displaystyle\int \dfrac{e^x - e^{-x}}{e^x + e^{-x}}\, dx = \ln\left(e^x + e^{-x}\right) + C$.

31. Let $u = \sqrt{x}$.

34. Let $u = \ln x$. Then $du = \dfrac{1}{x}\, dx$, and so

$$\int \frac{1}{x}(1 + \ln x)^{1/2}\, dx = \int (1 + u)^{1/2}\, du = \frac{2}{3}(1+u)^{3/2} + C = \frac{2}{3}(1 + \ln x)^{3/2} + C.$$

37. First, $dx = 2t\, dt$, and so $x = t^2 + C$. But $x(0) = 17$, and so $x(t) = t^2 + 17$.

40. $e^{-x}\, dx = dt$, so $-e^{-x} = C + t$. Now $-e^{-2} = C + 0$, and so

$$-e^{-x} = -e^{-2} + t;$$
$$e^{-x} = e^{-2} - t;$$
$$-x = \ln\left(e^{-2} - t\right);$$
$$x(t) = -\ln\left(e^{-2} - t\right).$$

43. $\dfrac{dx}{x} = \cos t\, dt$, so $\ln|x| = C_1 + \sin t$. Consequently $x(t) = Ce^{\sin t}$. Because $x(0) = \sqrt{2}$, $C = \sqrt{2}$. So $x(t) = \sqrt{2}\, e^{\sin t}$.

46. $\dfrac{dy}{dx} = 1 - \dfrac{1}{x}$ and $\dfrac{d^2y}{dx^2} = \dfrac{1}{x^2}$. The graph is concave upward for all x and there is a horizontal tangent at $(1,1)$. The function is decreasing on $(0,1)$ and increasing for $x > 1$. In addition, as $x \to 0^+$, $y(x) \to +\infty$ and $dy/dx \to -\infty$. The graph is shown on the right.

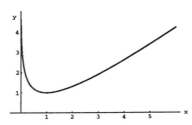

49. $\dfrac{dy}{dx} = \dfrac{1}{x^2}e^{-1/x}$ and $\dfrac{d^2y}{dx^2} = \dfrac{1-2x}{x^4}e^{-1/x}$.

52. At time t (in years), the amount to be repaid will be $1000e^{(0.1)t}$. So the profit on selling would be

$$P(t) = 800e^{\left(\frac{1}{2}\sqrt{t}\right)} - 1000e^{(t/10)}.$$
$$P'(t) = 200t^{-1/2}e^{\left(\frac{1}{2}\sqrt{t}\right)} - 100e^{(t/10)}.$$

$P'(t) = 0$ when $2e^{\left(\frac{1}{2}\sqrt{t}\right)} = t^{1/2}e^{(t/10)}$. The iteration

$$t \longleftarrow \left(2e^{\left(\frac{1}{2}\sqrt{t}\right)}e^{(-t/10)}\right)^2$$

yields $t \approx 11.7519$ years as the optimal time to cut and sell. The resulting profit would be \$1202.37.

58. Let $x(t)$ represent the position of the race car at time t and let $x(0) = 0$. Then the velocity of the car is $v(t) = x'(t)$. We want to find $v(0) = v_0$.

$$\frac{dv}{dt} = -kv,$$
$$v(t) = v_0e^{-kt}, \text{ as usual.}$$
$$x(t) = C - \frac{v_0}{k}\left(e^{-kt}\right).$$

Since $0 = x(0) = C - \dfrac{v_0}{k}$, $C = \dfrac{v_0}{k}$, hence $x(t) = \dfrac{v_0}{k}\left(1 - e^{-kt}\right)$.

When $t = 0$, $\dfrac{dv}{dt} = -2$, so $-kv_0 = -2$, hence $k = \dfrac{2}{v_0}$ and so $x(t) = \dfrac{(v_0)^2}{2}\left(1 - e^{-2t/v_0}\right)$.

$$\lim_{t \to \infty} v(t) = 0, \text{ so } \lim_{t \to \infty} x(t) = 1800.$$

$$1800 = \lim_{t \to \infty} \frac{(v_0)^2}{2}\left(1 - e^{-2t/v_0}\right) = \frac{(v_0)^2}{2}: \quad (v_0)^2 = 3600; \quad v_0 = 60.$$

The initial velocity of the car was 60 m/s.

61. Let $Q(t)$ denote the temperature at time t, with t in hours and $t = 0$ corresponding to 11:00 P.M., the time of the power failure. Let $A = 20$ be the room temperature. By Newton's law of cooling,

$$\frac{dQ}{dt} = k(A - Q), \text{ so that } \frac{1}{A - Q}\,dQ = k\,dt$$

where k is a positive constant. Next, $\ln(A - Q) = C_1 - kt$, so that $Q(t) = A - Ce^{-kt}$. Now $Q_0 = Q(0) = A - C$, so $C = A - Q_0$. Thus

$$Q(t) = A + (Q_0 - A)e^{-kt}.$$

We are given $Q_0 = -16$, $A = 20$, and $Q(7) = -10$. We must find the value $t = T$ at which $Q(T) = 0$. Now

$$Q(t) = 20 - 36e^{-kt};$$
$$Q(7) = -10 = 20 - 36e^{-7k};$$
$$30 = 36e^{-7k};$$
$$k = \tfrac{1}{7}\ln(1.2).$$

Next, $0 = Q(T) = 20 - 36e^{-kT}$; $20 = 36e^{-kT}$: Therefore $e^{kT} = 1.8$, and thus

$$T = \frac{1}{k}\ln(1.8) = \frac{7\ln(1.8)}{\ln(1.2)} \approx 22.5673.$$

The critical temperature will be reached about 22 hours and 34 minutes after the power goes off; that is, at about 9:34 P.M. on the following day. (The data used in this problem are those obtained during an actual incident of the sort described.)

64. Let S be the safe limit and let $R(t)$ be the radiation level at time t (in years).

$$R(t) = 10Se^{-kt}.$$

$R(\frac{1}{2}) = 95 = 105e^{-k/2}; \quad \frac{9}{10} = e^{-k/2}; \quad \frac{k}{2} = \ln \frac{10}{9}; \quad k = 2\ln \frac{10}{9}.$

$R(t) = S$ when $10Se^{-kt} = S$: $\quad e^{kt} = 10; \quad t = \dfrac{\ln 10}{k} = \dfrac{\ln 10}{2\ln\left(\frac{10}{9}\right)} \approx 10.927.$ Radiation will drop to a safe level in just under 11 years.

Chapter 8: Further Calculus of Transcendental Functions

Section 8.2

4. (a) $\sec^{-1}(1) = 0$ (b) $\sec^{-1}(-1) = \pi$

(c) $\sec^{-1}(2) = \pi/3$ (d) $\sec^{-1}(-\sqrt{2}) = 3\pi/4$

10. $f'(x) = \dfrac{x}{1+x^2} + \arctan x$

16. $f'(x) = \dfrac{1}{2x\sqrt{x-1}}$

22. $f'(x) = \dfrac{e^{\arcsin x}}{(1-x^2)^{1/2}}$

28. $\dfrac{dy}{dx} = -\left(1-y^2\right)^{1/2}\left(1-x^2\right)^{-1/2}$;   Tangent: $x\sqrt{3} + 3y = 2\sqrt{3}$

34. $\displaystyle\int_{-2}^{-2/\sqrt{3}} \dfrac{dx}{x\sqrt{x^2-1}} = \operatorname{arcsec}\left|\dfrac{-2}{\sqrt{3}}\right| - \operatorname{arcsec}|-2| = \dfrac{\pi}{6} - \dfrac{\pi}{3} = -\dfrac{\pi}{6}$

37. Suggestion: Let $u = 2x$.

40. Let $u = \frac{2}{3}x$, so that $2x = 3u$ and $dx = \frac{3}{2}\,du$. Then

$$\int \frac{dx}{x\left(4x^2-9\right)^{1/2}} = \int \frac{(3/2)\,du}{(3/2)u\left(9u^2-9\right)^{1/2}} = \int \frac{du}{3u\left(u^2-1\right)^{1/2}} = \tfrac{1}{3}\sec^{-1}|u| + C = \tfrac{1}{3}\sec^{-1}\left|\tfrac{2}{3}x\right| + C.$$

43. Suggestion: Let $u = \frac{1}{5}x^3$.

46. Suggestion: Let $u = \sec x$.

49. Suggestion: Let $u = \ln x$.

52. $\displaystyle\int_0^1 \frac{x^3}{1+x^4}\,dx = \left[\tfrac{1}{4}\ln\left(1+x^4\right)\right]_0^1 = \tfrac{1}{4}\ln 2.$

58. If $u = ax$, then $\sqrt{a^2-u^2} = \sqrt{a^2-a^2x^2} = a\sqrt{1-x^2}$ and $du = a\,dx$. Thus

$$\int \frac{1}{\sqrt{a^2-u^2}}\,dx = \int \frac{a}{a\sqrt{1-x^2}}\,dx = \arcsin x + C = \arcsin\left(\frac{u}{a}\right) + C.$$

The hypothesis that $a > 0$ is used in the first step, in which

$$\sqrt{a^2(1-x^2)} = |a|\sqrt{1-x^2} = a\sqrt{1-x^2}.$$

61. If $x > 1$, then

$$\frac{1}{x^2\sqrt{1-\dfrac{1}{x^2}}} = \frac{1}{x^2\sqrt{\dfrac{x^2-1}{x^2}}} = \frac{1}{\dfrac{x^2}{|x|}\sqrt{x^2-1}} = \frac{1}{x\sqrt{x^2-1}} = \frac{1}{|x|\sqrt{x^2-1}}.$$

The derivation requires slightly more care in the case $x < -1$ (so that $x < 0$), but generally follows the preceding line of argument.

64. (a) We begin with the identity

$$\tan(A+B) = \frac{\tan A + \tan B}{1 - \tan A \tan B}.$$

Let $x = \tan A$ and $y = \tan B$, and suppose that $xy < 1$. We will treat only the case in which x and y are both positive; the other cases are similar. In this case, the formula above shows that

$0 < A + B < \pi/2$, so it is valid to apply the inverse tangent function to each side of the above identity to obtain

$$A + B = \arctan \frac{x + y}{1 - xy},$$

and therefore

$$\arctan x + \arctan y = \arctan \frac{x + y}{1 - xy}.$$

(b i) $\quad \arctan \dfrac{(1/2) + (1/3)}{1 - (1/6)} = \arctan(1) = \pi/4.$

(b ii) $\quad \arctan \dfrac{(1/3) + (1/3)}{1 - (1/9)} + \arctan(1/7) = \arctan(3/4) + \arctan(1/7)$

$$= \arctan \frac{(3/4) + (1/7)}{1 - (3/28)} = \arctan \frac{25}{25} = \frac{\pi}{4}.$$

(b iii) $\arctan \dfrac{(120/119) - (1/239)}{1 + (120/119)(1/239)} = \arctan \dfrac{28561/28441}{28561/28441} = \dfrac{\pi}{4}.$

(b iv) $2 \arctan \dfrac{1}{5} = \arctan \dfrac{2/5}{1 - (1/25)} = \arctan \dfrac{10}{24} = \arctan \dfrac{5}{12};$

$$4 \arctan \frac{1}{5} = \arctan \frac{10/12}{1 - (25/144)} = \arctan \frac{120}{119};$$

the rest of Part (b *iv*) follows from Part (b *iii*)

70. Let y be the height of the elevator (measured upward from ground level) and let θ be the angle that your line of sight to the elevator makes with the horizontal ($\theta > 0$ if you are looking up, $\theta < 0$ if down). You're to maximize $\dfrac{d\theta}{dt}$ given $\dfrac{dy}{dt} = -25$.

$$\tan \theta = \frac{y - 100}{50}, \text{ so } \theta = \tan^{-1}\left(\frac{y - 100}{50} \right).$$

Therefore

$$\frac{d\theta}{dt} = \frac{d\theta}{dy} \cdot \frac{dy}{dt} = -25 \frac{1/50}{1 + ((y - 100)/50)^2} = \frac{(-25)(50)}{2500 + (y - 100)^2}.$$

To find the value of y that maximizes $f(y) = d\theta/dt$, we need only minimize the last denominator above: $y = 100$. Answer: The elevator has maximum apparent speed when it's at eye level.

76. The global maximum of $f(x)$ occurs at approximately $(8.333265, 1.334530)$.

Section 8.3

1. $\displaystyle \lim_{x \to 1} \frac{x - 1}{x^2 - 1} = \lim_{x \to 1} \frac{1}{2x} = \frac{1}{2}$. Of course, l'Hôpital's rule is not necessary here, but it may be applied.

4. $\displaystyle \lim_{x \to 0} \frac{e^{3x} - 1}{x} = \lim_{x \to 0} \frac{3e^{3x}}{1} = 3.$

7. $\displaystyle \lim_{x \to 1} \frac{x - 1}{\sin x} = \frac{0}{\sin 1} = 0$. Note that l'Hôpital's rule does not apply.

10. $\displaystyle \lim_{z \to \pi/2} \frac{1 + \cos 2z}{1 - \sin 2z} = \frac{1 + \cos \pi}{1 - \sin \pi} = \frac{0}{1} = 0.$

13. $\displaystyle \lim_{x \to \infty} \frac{\ln x}{x^{0.1}} = \lim_{x \to \infty} \frac{1/x}{(0.1)x^{-0.9}} = \lim_{x \to \infty} \frac{10}{x^{0.1}} = 0.$

16. $\displaystyle\lim_{t\to\infty} \frac{t^2+1}{t\ln t} = \lim_{t\to\infty} \frac{2t}{1+\ln t} = \lim_{t\to\infty} \frac{2}{1/t} = +\infty.$

It is also correct to say that this limit does not exist.

22. $\displaystyle\lim_{x\to 2} \frac{x^3-8}{x^4-16} = \frac{3}{8}.$

28. $\displaystyle\lim_{x\to\infty} \frac{\sqrt{x^3+x}}{\sqrt{2x^3-4}} = \frac{1}{\sqrt{2}}.$

34. $\displaystyle\lim_{x\to 0} \frac{e^{3x}-e^{-3x}}{2x} = 3.$

40. $\displaystyle\lim_{x\to\infty} \frac{\arctan 2x}{\arctan 3x} = 1.$

46. $\displaystyle\lim_{x\to\pi/4} \frac{1-\tan x}{4x-\pi} = -\frac{1}{2}.$

49. $\displaystyle\lim_{x\to 0} \frac{\sin^2 x}{x} = \lim_{x\to 0} (2\sin x \cos x) = 2\cdot 0\cdot 1 = 0,$ confirmed by the graph shown below.

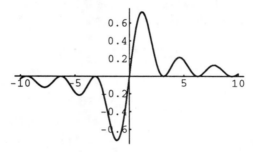

52. By l'Hôpital's rule, $\displaystyle\lim_{x\to\pi/2} \frac{\cos x}{2x-\pi} = \lim_{x\to\pi/2} \frac{-\sin x}{2} = -\frac{1}{2}.$ The graph (shown next) confirms this computation.

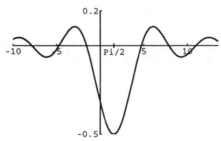

55. Here, l'Hôpital's rule yields $\displaystyle\lim_{x\to\infty} xe^{-x} = \lim_{x\to\infty} \frac{x}{e^x} = \lim_{x\to\infty} \frac{1}{e^x} = 0.$ Well, $x = 20$ isn't particularly "close" to $+\infty$, but the graph does suggest that the preceding computation is correct:

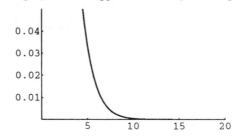

58. Two applications of l'Hôpital's rule yield $\lim\limits_{x\to\infty}\dfrac{x^2}{e^{2x}} = \lim\limits_{x\to\infty}\dfrac{2x}{2e^{2x}} = \lim\limits_{x\to\infty}\dfrac{2}{4e^{2x}} = 0$. The graph (next) confirms this result.

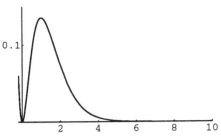

64. Given: $f(x) = x^{-k}\ln x$, where k is an arbitrary positive real number. Now

$$f'(x) = \frac{1 - k\ln x}{x^{k+1}}.$$

The sign of $f'(x)$ is that of $1 - k\ln x$, which is positive if $0 < x < e^{1/k}$ but negative if $x > e^{1/k}$. Hence the graph of f will have a single local maximum where $x = e^{1/k}$. Next,

$$f''(x) = \frac{k^2\ln x + k\ln x - 2k - 1}{x^{k+2}},$$

and $f''(x) = 0$ when $x = \exp\left(\dfrac{2k+1}{k^2+k}\right)$, so there is at most one inflection point on the graph of f. Moreover, if x is near zero, then $f''(x) < 0$, whereas $f''(x) > 0$ for large positive values of x. Hence there is exactly one inflection point. Finally, $\lim\limits_{x\to\infty}\dfrac{\ln x}{x^k} = \lim\limits_{x\to\infty}\dfrac{1}{kx^k} = 0$, and hence the (positive) x-axis is a horizontal asymptote.

67. By l'Hôpital's rule,

$$\lim_{h\to 0}\frac{f(x+h) - f(x-h)}{2h} = \lim_{h\to 0}\frac{f'(x+h) + f'(x-h)}{2} = \frac{2f'(x)}{2} = f'(x).$$

The continuity of f' is used to establish that $f'(x+h)$ and $f'(x-h)$ both approach $f'(x)$ as $h \to 0$. Note that the primes denote derivatives with respect to h.

Section 8.4

1. $\lim\limits_{x\to 0} x\cot x = \lim\limits_{x\to 0}\dfrac{x\cos x}{\sin x} = (1)(\cos 0) = 1.$

4. Let $u = \sin x$. Then $\lim\limits_{x\to 0+}(\sin x)(\ln\sin x) = \lim\limits_{u\to 0+} u\ln u = \lim\limits_{u\to 0+}\dfrac{\ln u}{1/u}$

$$= \lim_{u\to 0+}\frac{1/u}{-(1/u^2)} = \lim_{u\to 0+}(-u) = 0.$$

7. Let $u = 1/x$. Then $\lim\limits_{x\to\infty} x\left(e^{1/x} - 1\right) = \lim\limits_{u\to 0+}\dfrac{e^u - 1}{u} = \lim\limits_{u\to 0+}\dfrac{e^u}{1} = 1.$

10. Replace $\cos 3x$ with $\cos^3 x - 3\sin^2 x\cos x$; replace $\tan x$ with $(\sin x)/(\cos x)$.

Then $\lim\limits_{x\to\pi/2}(\tan x)(\cos 3x) = \lim\limits_{x\to\pi/2}(\sin x)\left(\cos^2 x - 3\sin^2 x\right) = -3.$

16. $\lim\limits_{x\to\infty}\left(\sqrt{x+1} - \sqrt{x}\right) = \lim\limits_{x\to\infty}\dfrac{x+1-x}{\sqrt{x+1}+\sqrt{x}} = \lim\limits_{x\to\infty}\dfrac{1}{\sqrt{x+1}+\sqrt{x}} = 0.$

22. Let $y = \left(x + \dfrac{1}{x}\right)^x$. Then

$$\ln y = x \ln \left(1 + \frac{1}{x}\right) = x \ln \left(\frac{x+1}{x}\right) = x \left[\ln(x+1) - \ln x\right] = \frac{\ln(x+1) - \ln x}{1/x}.$$

Now apply l'Hôpital's rule to find that

$$\lim_{x \to \infty} \ln y = \lim_{x \to \infty} \frac{\ln(x+1) - \ln x}{1/x} = \lim_{x \to \infty} \frac{x - (x+1)}{-(x+1)/x} = \lim_{x \to \infty} \frac{x}{x+1} = 1.$$

Therefore $y \to e$ as $x \to +\infty$.

25. The limit of the natural logarithm of the expression in question is $-1/6$. A calculator estimate of this limit is particularly misleading.

28. $\lim_{x \to 0^+} (\sin x)^{\sec x} = 0.$

34. Let $Q = x^5 - 3x^4 + 17$. Then $Q^{1/5} - x = \dfrac{Q - x^5}{Q^{4/5} + Q^{3/5}x + Q^{2/5}x^2 + Q^{1/5}x^3 + x^4}$. The numerator is just $-3x^4 + 17$. Now *carefully* divide each term in numerator and denominator by x^4; put this divisor within each radical in the denominator. Then let x increase without bound to see that the limit is $-3/5$.

37. (b) $f(x) \to 0$ as $x \to 0^+$ and $f(x) \to 1$ as $x \to +\infty$.

(c) The highest point on the graph of f is $(e, e^{2/e})$.

40. (b) Let $g(x) = \ln f(x) = \dfrac{\ln\left(1 + 1/x^2\right)}{1/x}$. Then $\lim_{x \to 0^+} g(x) = \lim_{x \to 0^+} \dfrac{2x}{1 + x^2} = 0$. Similarly, $g(x) \to 0$ as $x \to +\infty$. Therefore $f(x) \to 1$ as $x \to 0^+$ and also as $x \to +\infty$.

(c) We used Newton's method to determine that the highest point on the graph of f is approximately $(0.50498, 2.23612)$.

49. Graphically it is clear that $f(x) \to +\infty$ as $x \to 0^+$ and that $f(x) \to 1$ as $x \to +\infty$. The graph also shows a corner point at $(1, 0)$ much like the one on the graph of $y = |x|$, so there's a global minimum at $(1, 0)$. We were not able to accurately determine the inflection points graphically, so we applied Newton's method to $f''(x)$ to find their abscissas to be (approximately) 1.116390596 and 8.928007697. By "zooming" we found a local maximum near $(5.83120, 1.10215)$.

Section 8.5

1. $f'(x) = 3 \sinh(3x - 2)$

4. $f'(x) = -2e^{2x} \operatorname{sech} e^{2x} \tanh e^{2x}$

10. $f'(x) = \dfrac{\operatorname{sech}^2 x}{1 + \tanh^2 x}$

16. $\displaystyle\int \cosh^2 3u \, du = \int \tfrac{1}{2}(1 + \cosh 6u) \, du = \tfrac{1}{2}\left(u + \tfrac{1}{6}\sinh 6u\right) + C = \tfrac{1}{2}u + \tfrac{1}{6}\sinh 3u \cosh 3u + C$

19. If necessary, let $u = 2x$.

22. $\displaystyle\int \sinh^4 x \, dx = \int \left(\sinh^2 x\right)^2 dx = \int \tfrac{1}{4}(\cosh 2x - 1)^2 \, dx = \tfrac{1}{4}\int \left(\cosh^2 2x - 2\cosh 2x + 1\right) dx$

$$= \tfrac{1}{4}\int \left(\tfrac{1}{2}(\cosh 4x + 1) - 2\cosh 2x + 1\right) dx = \tfrac{1}{32}\sinh 4x - \tfrac{1}{4}\sinh 2x + \tfrac{3}{8}x + C$$

28. $\displaystyle\int \frac{e^x + e^{-x}}{e^x - e^{-x}} \, dx = \ln |e^x - e^{-x}| + C_1 = -\ln 2 + \ln |e^x - e^{-x}| + C = \ln |\sinh x| + C$

34. $f'(x) = -\dfrac{e^x}{e^x \, (1 + e^{2x})^{1/2}} = -\dfrac{1}{(1 + e^{2x})^{1/2}}$

40. Let $u = \frac{2}{3}y$. Then $y = \frac{3}{2}u$, $dy = \frac{3}{2}\,du$, and $4y^2 - 9 = 9\,(u^2 - 1)$. So

$$\int \frac{dy}{(4y^2 - 9)^{1/2}} = \int \frac{(3/2)\,du}{2\,(u^2 - 1)^{1/2}} = \tfrac{1}{2}\int \frac{du}{(u^2 - 1)^{1/2}} = \tfrac{1}{2}\cosh^{-1} u + C = \tfrac{1}{2}\cosh^{-1}\left(\frac{2y}{3}\right) + C.$$

43. Let $u = \frac{3}{2}x$.

46. Let $u = x^2$. Then $du = 2x\,dx$, $x\,dx = \frac{1}{2}\,du$, and thus

$$\int \frac{x\,dx}{(x^4 - 1)^{1/2}} = \tfrac{1}{2}\int \frac{du}{(u^2 - 1)^{1/2}} = \tfrac{1}{2}\cosh^{-1} u + C = \tfrac{1}{2}\cosh^{-1}\left(x^2\right) + C.$$

49. Proof:

$$\sinh x \cosh y + \cosh x \sinh y = \tfrac{1}{4}\left(e^x - e^{-x}\right)\left(e^y + e^{-y}\right) + \tfrac{1}{4}\left(e^x + e^{-x}\right)\left(e^y - e^{-y}\right)$$
$$= \tfrac{1}{4}\left(e^x e^y - e^{-x}e^y + e^x e^{-y} - e^{-x}e^{-y} + e^x e^y + e^{-x}e^y - e^x e^{-y} - e^{-x}e^{-y}\right)$$
$$= \tfrac{1}{4}\left(2e^x e^y - 2e^{-x}e^{-y}\right)$$
$$= \sinh(x + y).$$

52. If $x(t) = A\cosh kt + B\sinh kt$, then

$$x'(t) = kA\sinh kt + kB\cosh kt \text{ and}$$
$$x''(t) = k^2 A\cosh kt + k^2 B\sinh kt = k^2 x(t).$$

64. (a) We are to prove that

$$\coth^{-1} x = \tfrac{1}{2}\ln\left(\frac{x + 1}{x - 1}\right) \text{ if } |x| > 1.$$

By one of the formulas in the text, the left-hand side above has derivative

$$\frac{1}{1 - x^2}$$

The derivative of the right-hand side is

$$\left(\tfrac{1}{2}\right)\left(\frac{x - 1}{x + 1}\right)\left(\frac{x - 1 - x - 1}{(x - 1)^2}\right) = -\frac{1}{x^2 - 1} = \frac{1}{1 - x^2}.$$

Therefore

$$\coth^{-1} x = \frac{1}{2}\ln\frac{x + 1}{x - 1} + C \text{ for } |x| > 1.$$

(b) If $u = \coth^{-1}(2)$, then $\coth u = 2$. So

$$2 = \frac{e^u + e^{-u}}{e^u - e^{-u}};$$
$$2e^u - 2e^{-u} = e^u + e^{-u};$$
$$e^u = 3e^{-u}; \quad e^{2u} = 3; \quad 2u = \ln 3.$$

Therefore $u = \frac{1}{2}\ln 3$. Now substitute $x = 2$ in the result of Part (a):

$$\tfrac{1}{2}\ln 3 = \tfrac{1}{2}\ln\tfrac{3}{1} + C.$$

So $C = 0$ here as well. This establishes the desired result.

67. Note that $e^{-2x}\tanh x = \dfrac{1 - e^{-2x}}{1 + e^{2x}}$. Therefore $f(x) \to 0$ as $x \to +\infty$. We used Newton's method to find that the highest point on the graph of f is near $(0.440687, 0.171573)$.

70. **(a)** We let $f(k) = 30 + \dfrac{1}{k}(-1 + \cosh(100k)) - 50$ and solve the equation $f(k) = 0$ by Newton's method to discover that $k \approx 0.00394843545331550515$.

(b) We then let $y(x) = 30 + \dfrac{1}{k}(-1 + \cosh(kx))$, so that $\sqrt{1 + [(y'(x))^2} = \cosh(kx)$, and therefore the approximate length of the high-voltage line is

$$2 \int_0^{100} \cosh kx \, dx \approx 205.237 \quad \text{(ft)}.$$

Chapter 8 Miscellaneous

4. $g'(t) = \dfrac{e^t}{1 + e^{2t}}$

10. $f'(x) = -\dfrac{1}{x^2 + 1}$

16. $f'(x) = \dfrac{\sinh x}{\cosh x} = \tanh x$

22. $\displaystyle\int \dfrac{dx}{1 + 4x^2} = \tfrac{1}{2}\arctan 2x + C$

28. Let $u = \tfrac{2}{3}x$; $\displaystyle\int \dfrac{1}{9 + 4x^2}\, dx = \tfrac{3}{2}\int \dfrac{1}{9 + 9u^2}\, du = \tfrac{1}{6}\arctan u + C = \tfrac{1}{6}\arctan\left(\dfrac{2x}{3}\right) + C$.

31. Let $u = 2x$; $x = \tfrac{1}{2}u$ and $dx = \tfrac{1}{2}\,du$. Then

$$\int \dfrac{1}{x\sqrt{4x^2 - 1}}\, dx = \int \dfrac{du}{u\sqrt{u^2 - 1}} = \sec^{-1}|2x| + C.$$

34. $\displaystyle\int x^2 \cosh x^3 \, dx = \tfrac{1}{3}\sinh x^3 + C$.

40. Let $u = x^2$; then $du = 2x\, dx$, and the integral becomes $\dfrac{1}{2}\displaystyle\int \dfrac{du}{\sqrt{u^2 + 1}} = \tfrac{1}{2}\sinh^{-1}\left(x^2\right) + C$.

43. $\displaystyle\lim_{x \to \pi} \dfrac{1 + \cos x}{(x - \pi)^2} = \lim_{x \to \pi} \dfrac{-\sin x}{2(x - \pi)} = \lim_{x \to \pi} \dfrac{-\cos x}{2} = \tfrac{1}{2}$.

46. $\displaystyle\lim_{x \to \infty} \dfrac{\ln(\ln x)}{\ln x} = \lim_{x \to \infty} \dfrac{\left(\dfrac{1}{x \ln x}\right)}{\left(\dfrac{1}{x}\right)} = \lim_{x \to \infty} \dfrac{1}{\ln x} = 0$.

49. $\displaystyle\lim_{x \to 0}\left(\dfrac{1}{x^2} - \dfrac{1}{1 - \cos x}\right) = \lim_{x \to 0} \dfrac{1 - \cos x - x^2}{x^2(1 - \cos x)} = \lim_{x \to 0} \dfrac{\sin x - 2x}{2x(1 - \cos x) + x^2 \sin x}$

$$= \lim_{x \to 0} \dfrac{\left(\dfrac{\sin x}{x} - 2\right)}{2(1 - \cos x) + x \sin x} = -\infty.$$

52. $\displaystyle\lim_{x \to \infty} \ln\left(x^{1/x}\right) = \lim_{x \to \infty} \dfrac{\ln x}{x} = 0$. Therefore $\displaystyle\lim_{x \to \infty} x^{1/x} = 1$.

55. One of our most challenging problems. First let $u = 1/x$. It is then sufficient to evaluate

$$L = \lim_{u \to 0+} \dfrac{(1 + u)^{1/u} - e}{u}.$$

Apply l'Hôpital's rule once:

$$L = \lim_{u \to 0+} (1 + u)^{1/u} \left(\dfrac{u - (1 + u)\ln(1 + u)}{u^2(1 + u)}\right).$$

Now apply the product rule for limits!

$$L = e \lim_{u \to 0+} \frac{u - (1 + u) \ln(1 + u)}{u^2(1 + u)}.$$

Finally apply l'Hôpital's rule twice to the limit that remains. The answer in the text follows without any more difficulties.

58. By the method of cylindrical shells,

$$V = \int_0^1 2\pi x \, \frac{1}{\sqrt{x^4 + 1}} \, dx.$$

Let $u = x^2$. Then $du = 2x \, dx$, and we obtain

$$V = \int_0^1 \frac{\pi}{\sqrt{u^2 + 1}} \, du = \pi \left(\sinh^{-1}(1) - \sinh^{-1}(0) \right) = \pi \sinh^{-1}(1) = \pi \ln \left(1 + \sqrt{2} \right) \approx 2.7689.$$

61. It's clear from the sketch on the right that the least positive solution is larger than $3\pi/2$ and smaller than 2π. So we use the iteration of Newton's method with initial estimate $x_0 = 5$. The formula is

$$x \longleftarrow x - \frac{\cos x \, \cosh x - 1}{\cos x \, \sinh x - \sin x \, \cosh x}.$$

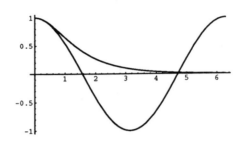

Here are the (rounded) results:

$$x_0 = 5.0$$
$$x_1 = 4.782556376$$
$$x_2 = 4.732575187$$
$$x_3 = 4.730047035$$
$$x_4 = 4.730040745 = x_5.$$

Answer: The least positive solution of $\cos x \, \cosh x = 1$ is approximately 4.730040745.

Chapter 9: Techniques of Integration

Section 9.2

1. Let $u = 2 - 3x$ and apply Formula (1) of Fig. 9.1.

4. Let $u = 5 + 2t^2$. $\int \dfrac{5t}{5 + 2t^2}\, dt = \dfrac{5}{4} \int \dfrac{du}{u} = \dfrac{5}{4} \ln |u| + C = \dfrac{5}{4} \ln\left(5 + 2t^2\right) + C.$

7. Let $u = \sqrt{y}$.

10. Let $u = 4 + \cos 2x$. $\int \dfrac{\sin 2x}{4 + \cos 2x}\, dx = -\dfrac{1}{2} \int \dfrac{du}{u} = -\dfrac{1}{2} \ln |u| + C = -\dfrac{1}{2} \ln(4 + \cos 2x) + C.$

13. Let $u = \ln t$.

16. Let $u = 1 + e^{2x}$. $\int \dfrac{e^{2x}}{1 + e^{2x}}\, dx = \dfrac{1}{2} \int \dfrac{du}{u} = \dfrac{1}{2} \ln |u| + C = \dfrac{1}{2} \ln\left(1 + e^{2x}\right) + C$

19. Let $u = x^2$.

22. Let $u = 2t$. $\int \dfrac{1}{1 + 4t^2}\, dt = \dfrac{1}{2} \int \dfrac{du}{1 + u^2} = \dfrac{1}{2} \arctan u + C = \dfrac{1}{2} \arctan 2t + C.$

25. Let $v = 1 + \sqrt{x}$. $\int \dfrac{(1 + \sqrt{x})^4}{\sqrt{x}}\, dx = 2 \int v^4\, dv = \dfrac{2}{5} v^5 + C = \dfrac{2}{5} \left(1 + \sqrt{x}\right)^5 + C.$

28. Let $u = 1 + \sec 2x$. $\int \dfrac{\sec 2x \tan 2x}{(1 + \sec 2x)^{3/2}}\, dx = \int \dfrac{1}{2} u^{-3/2}\, du = -u^{-1/2} + C = -\dfrac{1}{\sqrt{1 + \sec 2x}} + C.$

31. $\displaystyle\int x^2\sqrt{x - 2}\, dx = \int (u + 2)^2 u^{1/2}\, du = \int \left(u^{5/2} + 4u^{3/2} + 4u^{1/2}\right)\, du$

$= \dfrac{2}{7} u^{7/2} + \dfrac{8}{5} u^{5/2} + \dfrac{8}{3} u^{3/2} + C = \dfrac{2}{7}(x - 2)^{7/2} + \dfrac{8}{5}(x - 2)^{5/2} + \dfrac{8}{3}(x - 2)^{3/2} + C$

$= (x - 2)^{3/2} \left(\dfrac{2}{7}(x - 2)^2 + \dfrac{8}{5}(x - 2) + \dfrac{8}{3}\right) + C$

$= \dfrac{1}{105}(x - 2)^{3/2} \left(30 \left(x^2 - 4x + 4\right) + 168(x - 2) + 280\right) + C$

$= \dfrac{2}{105}(x - 2)^{3/2} \left(15x^2 + 24x + 32\right) + C.$

34. $\displaystyle\int x\sqrt[3]{x - 1}\, dx = \int (u + 1) u^{1/3}\, du = \int \left(u^{4/3} + u^{1/3}\right)\, du$

$= \dfrac{3}{7} u^{7/3} + \dfrac{3}{4} u^{4/3} + C = \left(3u^{4/3}\right) \cdot \left(\dfrac{4u + 7}{28}\right) + C$

$= \left(\dfrac{3(x - 1)^{4/3}}{28}\right) \cdot (4x + 3) + C = \dfrac{3}{28}(x - 1)^{1/3} \left(4x^2 - x - 3\right) + C.$

37. With $a = 10$, $u = 3x$, and $dx = \dfrac{1}{3} du$, we get

$\displaystyle\int \dfrac{1}{100 - 9x^2}\, dx = \dfrac{1}{3} \int \dfrac{1}{a^2 - u^2}\, du = \dfrac{1}{6a} \ln \left|\dfrac{u + a}{u - a}\right| + C = \dfrac{1}{60} \ln \left|\dfrac{3x + 10}{3x - 10}\right| + C.$

40. With $a = 3$, $u = 4x$, and $dx = \dfrac{1}{4} du$, we obtain

$\displaystyle\int \dfrac{dx}{\sqrt{16x^2 + 9}} = \dfrac{1}{4} \int \dfrac{du}{(u^2 + a^2)^{1/2}} = \dfrac{1}{4} \ln \left|u + (u^2 + a^2)^{1/2}\right| + C = \dfrac{1}{4} \ln \left|4x + (16x^2 + 9)^{1/2}\right| + C.$

43. With $a = 5$, $u = 4x$, and $dx = \dfrac{1}{4} du$, we get

$\displaystyle\int x^2\sqrt{25 - 16x^2}\, dx = \dfrac{1}{64} \int u^2 \left(a^2 - u^2\right)^{1/2}\, du$

$= \dfrac{1}{64} \left(\dfrac{u}{8} \left(2u^2 - a^2\right) \left(a^2 - u^2\right)^{1/2} + \dfrac{1}{8} a^4 \arcsin \dfrac{u}{a}\right) + C$

$= \dfrac{x}{128} \left(32x^2 - 25\right) \left(25 - 16x^2\right)^{1/2} + \dfrac{625}{512} \arcsin \dfrac{4x}{5} + C.$

46. Let $u = \sin x$, $du = \cos x\, dx$. Use Formula (50) of the endpapers to obtain

$$\int \frac{\cos x}{(\sin^2 x)\sqrt{1+\sin^2 x}}\, dx = \int \frac{1}{u^2\,(1+u^2)^{1/2}}\, du = -\frac{1}{u}\,(u^2+1)^{1/2} + C = -\frac{(1+\sin^2 x)^{1/2}}{\sin x} + C$$

49. The substitution $u = \ln x$ leads to

$$\int \frac{(\ln x)^2}{x}\,\sqrt{1+(\ln x)^2}\, dx = \int u^2\,(1+u^2)^{1/2}\, du \qquad \text{(which by Formula (48))}$$

$$= \frac{1}{8}\Big[(\ln x)\,(2(\ln x)^2+1)\,((\ln x)^2+1)^{1/2} - \ln\big|(\ln x) + ((\ln x)^2+1)^{1/2}\big|\Big] + C.$$

52. $\displaystyle\int \frac{dx}{x^2+4x+5} = \int \frac{dx}{(x+2)^2+1} = \tan^{-1}(x+2) + C.$

Section 9.3

1.
$$\begin{bmatrix} u = x & dv = e^{2x}\, dx \\ du = dx & v = \frac{1}{2}e^{2x} \end{bmatrix}$$
$$\int xe^{2x}\, dx = \frac{1}{2}xe^{2x} - \int \frac{1}{2}e^{2x}\, dx = \frac{1}{2}xe^{2x} - \frac{1}{4}e^{2x} + C.$$

4.
$$\begin{bmatrix} u = t^2 & dv = \sin t\, dt \\ du = 2t\, dt & v = -\cos t \end{bmatrix}$$
$$\int t^2 \sin t\, dt = -t^2 \cos t + 2\int t \cos t\, dt.$$

7.
$$\begin{bmatrix} u = \ln x & dv = x^3\, dx \\ du = \frac{1}{x}\, dx & v = \frac{1}{4}x^4 \end{bmatrix}$$
$$\int x^3 \ln x\, dx = \frac{1}{4}x^4 \ln x - \frac{1}{16}x^4 + C.$$

10.
$$\begin{bmatrix} u = \ln x & dv = \frac{1}{x^2}\, dx \\ du = \frac{1}{x}\, dx & v = -\frac{1}{x} \end{bmatrix}$$
$$\int \frac{\ln x}{x^2}\, dx = -\frac{1}{x}\ln x + \int \frac{1}{x^2}\, dx = -\frac{1}{x}\ln x - \frac{1}{x} + C.$$

13. Choose $u = (\ln t)^2$ and $dv = dt$. The necessity of integrating $\ln t$ then arises; a second integration by parts is necessary, and works well with the choices $u = \ln t$, $dv = dt$.

16.
$$\begin{bmatrix} u = x^2 & dv = x\,(1-x^2)^{1/2}\, dx \\ du = 2x\, dx & v = -\frac{1}{3}\,(1-x^2)^{3/2} \end{bmatrix} \int x^3\sqrt{1-x^2}\, dx = -\frac{1}{3}x^2\,(1-x^2)^{3/2} - \frac{2}{15}\,(1-x^2)^{5/2} + C.$$

19. Choose $u = \csc\theta$, $dv = \csc^2\theta\, d\theta$. Replace $\cot^2\theta$ with $\csc^2\theta - z$ in the resulting integral to obtain

$$I = -\csc\theta\,\cot\theta + \int \csc\theta\, d\theta - I$$

where I is the original integral. Solve for I and use Formula 15 of the endpapers to obtain the answer.

22. Choose $u = \ln\left(1+x^2\right)$, $dv = dx$. Then

$$\int \ln\left(1+x^2\right)\, dx = x\ln\left(1+x^2\right) - 2\int \frac{x^2}{x^2+1}\, dx$$
$$= x\ln\left(1+x^2\right) - 2\int \frac{x^2+1}{x^2+1}\, dx + 2\int \frac{dx}{x^2+1}$$
$$= x\ln\left(1+x^2\right) - 2x + 2\tan^{-1}x + C.$$

25. Choose $u = \tan^{-1}\sqrt{x}$ and $dv = dx$. To simplify the resulting computations, choose $v = x + 1$.

28. Choose $u = \tan^{-1}x$, $dv = x\, dx$. Then $du = \dfrac{dx}{1+x^2}$; we choose $v = \frac{1}{2}x^2 + \frac{1}{2} = \frac{1}{2}\left(x^2+1\right)$ for the purpose of providing a fortuitous cancellation:

$$\int x\tan^{-1}x\, dx = \frac{1}{2}\left(x^2+1\right)\tan^{-1}x - \frac{1}{2}\int \frac{x^2+1}{1+x^2}\, dx = \frac{1}{2}\left(x^2+1\right)\tan^{-1}x - \frac{1}{2}x + C.$$

This sly trick—a clever choice for v—is rarely useful. Try it with the integral of $\ln(x+1)$.

31. Choose $u = \ln x$ and $dv = x^{-3/2}\, dx$. Then

$$\int \frac{\ln x}{x^{3/2}}\, dx = -2x^{-1/2}\ln x + 2\int x^{-3/2}\, dx$$
$$= -2x^{-1/2}\ln x - 4x^{-1/2} + C$$
$$= -\frac{2}{\sqrt{x}}\,(2 + \ln x) + C.$$

34. First method:

$$\int e^x \cosh x\, dx = \tfrac{1}{2}\int \left(e^{2x} + 1\right)\, dx = \tfrac{1}{4}e^{2x} + \tfrac{1}{2}x + C_1$$
$$= \tfrac{1}{4}\left(e^{2x} + 1\right) - \tfrac{1}{4} + \tfrac{1}{2}x + C_1 = \tfrac{1}{4}e^x\left(e^x + e^{-x}\right) + \tfrac{1}{2}x + C$$
$$= \tfrac{1}{2}e^x \cosh x + \tfrac{1}{2}x + C.$$

Second method: Presented because no integration by parts is used in the first method, although what follows is somewhat artificial.

$$\begin{bmatrix} u = e^x & dv = \cosh x\, dx \\ du = e^x\, dx & v = \sinh x \end{bmatrix} \qquad J = \int e^x \cosh x\, dx = e^x \sinh x - \int e^x \sinh x\, dx.$$

Now $e^x \sinh x = \tfrac{1}{2}\left(e^{2x} - 1\right) = \tfrac{1}{2}\left(e^{2x} + 1\right) - 1 = e^x \cosh x - 1$. Therefore

$$J = e^x \sinh x - J + \int 1\, dx; \ \text{it follows that} \int e^x \cosh x\, dx = \tfrac{1}{2}e^x \sinh x + \tfrac{1}{2}x + C.$$

37. Let $t = \sqrt{x}$, so that $x = t^2$ and $dx = 2t\, dt$. Thus

$$I = \int \exp\left(-\sqrt{x}\right)\, dx = \int 2t \exp(-t)\, dt.$$

Now let $u = 2t$ and $dv = \exp(-t)\, dt$. Then $du = 2\, dt$ and $v = -\exp(-t)$. Hence

$$I = -2t \exp(-t) + \int 2 \exp(-t)\, dt = -2t \exp(-t) - 2 \exp(-t) + C.$$

Therefore

$$I = -2\sqrt{x} \exp\left(-\sqrt{x}\right) - 2 \exp\left(-\sqrt{x}\right) + C.$$

40. $\displaystyle\int_0^\pi 2\pi x \sin x\, dx = 2\pi^2.$

43. The curves intersect at the point (a, b) in the first quadrant for which $a \approx 0.824132312$. The volume is

$$\int_0^a 2\pi x \left((\cos x) - x^2\right)\, dx \approx 1.06027.$$

46. $\displaystyle\int 2x \arctan x\, dx = \left(x^2 + 1\right) \tan^{-1} x - x + C.$

One may add any convenient constant to v. Here's why: Let K be any constant. Then

$$\int u\, d(v + K) = u(v + K) - \int (v + K)\, du$$
$$= uv + uK - \int v\, du - \int K\, du$$
$$= uv + uK - \int v\, du - Ku$$
$$= uv - \int v\, du = \int u\, dv.$$

49. Let $u = x^n$ and $dv = e^x \, dx$. Then $du = nx^{n-1} \, dx$ and $v = e^x$. The desired reduction formula is an immediate consequence.

52. $\left[\begin{array}{cc} u = x^n & dv = \cos x \, dx \\ du = nx^{n-1} \, dx & v = \sin x \end{array} \right]$ $\qquad \displaystyle\int x^n \cos x \, dx = x^n \sin x - n \int x^{n-1} \sin x \, dx.$

55. $\displaystyle\int_0^1 x^3 e^x \, dx = \left[x^3 e^x \right]_0^1 - 3 \int_0^1 x^2 e^x \, dx$

$\displaystyle = e - 3 \left[x^2 e^x \right]_0^1 + 6 \int_0^1 x e^x \, dx$

$\displaystyle = e - 3e + 6 \left[x e^x \right]_0^1 - 6 \left[e^x \right]_0^1$

$\displaystyle = e - 3e + 6e - 6e + 6 = 6 - 2e \approx 0.563436.$

58. Let $\displaystyle S_{2n} = \int_0^{\pi/2} \sin^{2n} x \, dx$. By the solution of Problem 45, $\displaystyle S_{2n} = \frac{2n - 1}{2n} S_{2n-2}$.

Repeated application of this formula yields

$$S_{2n} = \left(\frac{2n - 1}{2n} \right) \cdot \left(\frac{2n - 3}{2n - 2} \right) S_{2n-4}$$

$$= \left(\frac{2n - 1}{2n} \right) \cdot \left(\frac{2n - 3}{2n - 2} \right) \cdot \left(\frac{2n - 5}{2n - 4} \right) S_{2n-6} = \cdots$$

$$= \left(\frac{2n - 1}{2n} \right) \cdot \left(\frac{2n - 3}{2n - 2} \right) \cdot \left(\frac{2n - 5}{2n - 4} \right) \cdots \frac{3}{4} \cdot \frac{1}{2} \int_0^{\pi/2} 1 \, dx$$

$$= \left(\frac{2n - 1}{2n} \right) \cdot \left(\frac{2n - 3}{2n - 2} \right) \cdots \frac{5}{6} \cdot \frac{3}{4} \cdot \frac{1}{2} \cdot \frac{\pi}{2}.$$

The other derivation is similar.

64. (a) Area:

$$A = \int_0^\pi \tfrac{1}{2} x^2 \sin x \, dx = \left[\tfrac{1}{2} \left(-x^2 \cos x + 2 \int x \cos x \, dx \right) \right]_0^\pi$$

$$= \left[-\tfrac{1}{2} x^2 \cos x + x \sin x - \int \sin x \, dx \right]_0^\pi$$

$$= \left[-\tfrac{1}{2} x^2 \cos x + x \sin x + \cos x \right]_0^\pi$$

$$= \tfrac{1}{2} \pi^2 - 1 - 1 = \frac{\pi^2 - 4}{2}.$$

(b) Volume:

$$V = \int_0^\pi 2\pi x \cdot \tfrac{1}{2} x^2 \sin x \, dx = \pi \int_0^\pi x^3 \sin x \, dx$$

$$= \pi \left[-x^3 \cos x + 3 \int x^2 \cos x \, dx \right]_0^\pi$$

$$= \pi \left[-x^3 \cos x + 3 \left(x^2 \sin x - 2 \int x \sin x \, dx \right) \right]_0^\pi$$

$$= \pi \left[-x^3 \cos x + 3x^2 \sin x - 6 \left(-x \cos x + \sin x \right) \right]_0^\pi$$

$$= \pi \left[-x^3 \cos x + 3x^2 \sin x + 6x \cos x - 6 \sin x \right]_0^\pi$$

$$= \pi \left(\pi^3 - 6\pi \right) = \pi^4 - 6\pi^2 = \pi^2 \left(\pi^2 - 6 \right).$$

Section 9.4

1. $\int \sin^2 2x \, dx = \int \frac{1 - \cos 4x}{2} \, dx = \frac{x}{2} - \frac{1}{8} \sin 4x + C.$

4. $\int \tan^2 \frac{x}{2} \, dx = \int \left(\sec^2 \frac{x}{2} - 1 \right) \, dx = \left(2 \tan \frac{x}{2} \right) - x + C.$

7. $\int \sec 3x \, dx = \frac{1}{3} \ln |\sec 3x + \tan 3x| + C.$

10. $\int \sin^2 x \cot^2 x \, dx = \int \cos^2 x \, dx = \int \frac{1 + \cos 2x}{2} \, dx = \frac{1}{2}x + \frac{1}{4} \sin 2x + C.$

13. $\int \sin^2 \theta \cos^3 \theta \, d\theta = \int (\sin^2 \theta)(1 - \sin^2 \theta)(\cos \theta) \, d\theta$

$$= \int (\sin^2 \theta \cos \theta - \sin^4 \theta \cos \theta) \, d\theta = \frac{1}{3} \sin^3 \theta - \frac{1}{5} \sin^5 \theta + C.$$

16. $\int \frac{\sin t}{\cos^3 t} \, dt = \frac{1}{2 \cos^2 t} + C.$

19. $\int \sin^5 2z \cos^2 2z \, dz = \int (\sin 2z)(1 - \cos^2 2z)^2 \cos^2 2z \, dz$

$$= \int (\cos^6 2z \sin 2z - 2 \cos^4 2z \sin 2z + \cos^2 2z \sin 2z) \, dz$$

$$= -\frac{1}{14} \cos^7 2z + \frac{1}{5} \cos^5 2z - \frac{1}{6} \cos^3 2z + C.$$

22. $\cos^6 u = \left(\cos^2 u \right)^3 = \frac{1}{8}(1 + \cos 2u)^3 = \frac{1}{8}\left(1 + 3\cos 2u + 3\cos^2 2u + \cos^3 2u \right)$

$$= \frac{1}{8}\left(1 + 3 \cos 2u + \frac{3}{2}(1 + \cos 4u) + (1 - \sin^2 2u)(\cos 2u)\right).$$

Thus the antiderivative—simplified—is $\frac{5}{16}\theta + \frac{1}{16} \sin 8\theta + \frac{3}{256} \sin 16\theta - \frac{1}{192} \sin^3 8\theta + C.$

25. $\int \cot^3 2x \, dx = \int (\cot 2x)(\csc^2 2x - 1) \, dx.$

28. $\int \cot^3 x \csc^2 x \, dx = -\frac{1}{4} \cot^4 x + C.$

31. $\int \frac{\tan^3 \theta}{\sec^4 \theta} \, d\theta = \int (\sec \theta)^{-4} (\sec^2 \theta - 1)(\tan \theta) \, d\theta$

$$= \int ((\sec \theta)^{-3} \sec \theta \tan \theta - (\sec \theta)^{-5} \sec \theta \tan \theta) \, d\theta.$$

34. $\int \frac{1}{\cos^4 2x} \, dx = \int \sec^4 2x \, dx = \int (\tan^2 2x + 1) \sec^2 2x \, dx = \frac{1}{6} \tan^3 2x + \frac{1}{2} \tan 2x + C.$

37. $\int \cos^3 5t \, dt = \int (1 - \sin^2 5t) \cos 5t \, dt = \int (\cos 5t - \sin^2 5t \cos 5t) \, dt.$

40. $\int \tan^2 2t \sec^4 2t \, dt = \int (\tan^2 2t)(\tan^2 2t + 1) \sec^2 2t \, dt = \int (\tan^4 2t \sec^2 2t + \tan^2 2t \sec^2 2t) \, dt$

$$= \frac{1}{10} \tan^5 2t + \frac{1}{6} \tan^3 2t + C.$$

43. $\frac{\tan x}{\sec x} = \sin x$ and $\frac{\sin x}{\sec x} = \sin x \cos x.$

46. The area is $\int_{-\pi/4}^{\pi/4} (\cos^2 x - \sin^2 x) \, dx = \int_{-\pi/4}^{\pi/4} \cos 2x \, dx = \left[\frac{1}{2} \sin 2x \right]_{-\pi/4}^{\pi/4} = 1.$

49. 0

52. The volume is

$$\int_{-\pi/4}^{\pi/4} \left(\pi \cos^4 x - \pi \sin^4 x \right) \, dx = \pi \int_{-\pi/4}^{\pi/4} \left(\cos^2 x + \sin^2 x \right) \left(\cos^2 x - \sin^2 x \right) \, dx$$

$$= \pi \int_{-\pi/4}^{\pi/4} \cos 2x \, dx = \pi \left[\tfrac{1}{2} \sin 2x \right]_{-\pi/4}^{\pi/4} = \pi.$$

55. **(a)** The area of R is $\displaystyle\int_0^{\pi/4} \left(\sec^2 x - \tan^2 x \right) \, dx = \int_0^{\pi/4} 1 \, dx = \frac{\pi}{4}.$

(b) The volume of revolution is

$$\int_0^{\pi/4} \left(\pi \sec^4 x - \pi \tan^4 x \right) \, dx = \pi \int_0^{\pi/4} \left(\sec^2 x + \tan^2 x \right) \left(\sec^2 x - \tan^2 x \right) \, dx$$

$$= \pi \int_0^{\pi/4} \left(2 \sec^2 x - 1 \right) \, dx = \pi \left[2 \tan x - x \right]_0^{\pi/4}$$

$$= \frac{\pi(8 - \pi)}{4} \approx 3.815784207.$$

58. **(1)** $\displaystyle\int \cot^3 x \, dx = \int (\cot x) \left(\csc^2 x - 1 \right) \, dx = -\tfrac{1}{2} \cot^2 x - \ln |\sin x| + C_1;$

(2) $\displaystyle\int \cot^3 x \, dx = \int \frac{\cos^3 x}{\sin^3 x} \, dx = \int \frac{(\cos x)\left(1 - \sin^2 x\right)}{\sin^3 x} \, dx = \int \left(\frac{\cos x}{\sin^3 x} - \frac{\cos x}{\sin x} \right) \, dx$

$$= -\frac{1}{2 \sin^2 x} - \ln |\sin x| + C_2.$$

The difference in the two antiderivatives is $\tfrac{1}{2} + C_1 - C_2 = C$ (a constant). Therefore the two results are equivalent.

61. $\cos x \cos 4x = \tfrac{1}{2}(\cos 3x + \cos 5x).$

Section 9.5

1. $\displaystyle\int \frac{x^2}{x+1} \, dx = \int \left(x - 1 + \frac{1}{x+1} \right) dx = \tfrac{1}{2}x^2 - x + \ln |x+1| + C.$

4. $\displaystyle\int \frac{x}{x^2 + 4x} \, dx = \int \frac{1}{x+4} \, dx = \ln |x + 4| + C.$

7. $\displaystyle\frac{1}{x^3 + 4x} = \frac{1}{x\left(x^2 + 4\right)} = \frac{A}{x} + \frac{Bx + C}{x^2 + 4}$ yields $A = \tfrac{1}{4}$, $B = -\tfrac{1}{4}$, and $C = 0$.

So $\displaystyle\int \frac{1}{x^3 + 4x} \, dx = \tfrac{1}{4} \ln |x| - \tfrac{1}{8} \ln \left(x^2 + 4\right) + C.$

10. $\displaystyle\frac{1}{\left(x^2 + 1\right)\left(x^2 + 4\right)} = \frac{Ax + B}{x^2 + 1} + \frac{Cx + D}{x^2 + 4}$ yields $A = C = 0$, $B = \tfrac{1}{3}$, and $D = -\tfrac{1}{3}$.

Thus $\displaystyle\int \frac{1}{\left(x^2 + 1\right)\left(x^2 + 4\right)} \, dx = \tfrac{1}{3} \arctan x - \tfrac{1}{6} \arctan \left(\frac{x}{2} \right) + C.$

13. Rewrite the integrand as $1 - \dfrac{1}{(x+1)^2}.$

16. After division:

$$\int \frac{x^4}{x^2 + 4x + 4} \, dx = \int \left(x^2 - 4x + 12 - (16)\frac{2x + 3}{(x+2)^2} \right) dx$$

$$= \tfrac{1}{3}x^3 - 2x^2 + 12x - 16 \int \frac{2x + 4 - 1}{(x + 2)^2} \, dx$$

$$= \tfrac{1}{3}x^3 - 2x^2 + 12x - 16 \int \left(\frac{2}{x + 2} - \frac{1}{(x + 2)^2} \right) dx$$

$$= \tfrac{1}{3}x^3 - 2x^2 + 12x - \frac{16}{x + 2} - 32 \ln |x + 2| + C.$$

19. $\dfrac{x^2 + 1}{x^3 + 2x^2 + x} = \dfrac{1}{x} + \dfrac{0}{x + 1} - \dfrac{2}{(x + 1)^2}.$

22. $\displaystyle\int \frac{2x^2 + 3}{x^4 - 2x^2 + 1} \, dx = \int \left(-\tfrac{1}{4} \left(\frac{1}{x - 1} \right) + \tfrac{5}{4} \left(\frac{1}{(x - 1)^2} \right) + \tfrac{1}{4} \left(\frac{1}{x + 1} \right) + \tfrac{5}{4} \left(\frac{1}{(x + 1)^2} \right) \right) dx$

$$= \tfrac{1}{4} \ln \left| \frac{x + 1}{x - 1} \right| - \frac{5x}{2(x^2 - 1)} + C.$$

25. $\dfrac{1}{x^3 + x} = \dfrac{A}{x} + \dfrac{Bx + C}{x^2 + 1}$ leads to $A = 1, B = -1, C = 0.$

28. $\dfrac{4x^4 + x + 1}{x^5 + x^4} = \dfrac{A}{x} + \dfrac{B}{x^2} + \dfrac{C}{x^3} + \dfrac{D}{x^4} + \dfrac{E}{x + 1}$ yields $A = B = C = 0, D = 1,$ and $E = 4.$

So $\displaystyle\int \frac{4x^4 + x + 1}{x^5 + x^4} \, dx = \int \left(\frac{1}{x^4} + \frac{4}{x + 1} \right) dx = -\tfrac{1}{3}x^{-3} + 4 \ln |x + 1| + C.$

31. $\dfrac{x^2 - 10}{2x^4 + 9x^2 + 4} = \dfrac{Ax + B}{2x^2 + 1} + \dfrac{Cx + D}{x^2 + 4}$ yields $A = C = 0, B = -3,$ and $D = 2.$

34. $\displaystyle\int \frac{x^2 + 4}{(x^2 + 1)^2 (x^2 + 2)} \, dx = \int \left(-\frac{2}{x^2 + 1} + \frac{3}{(x^2 + 1)^2} + \frac{2}{x^2 + 2} \right) dx.$

The first term in the integrand presents no problem, and by Formula 17 of the endpapers the third yields

$$\sqrt{2} \arctan \left(\tfrac{1}{2}x\sqrt{2} \right) + C_3.$$

For the second term, we use the substitution $x = \tan z$. Then

$$\int \frac{3}{(x^2 + 1)^2} \, dx = 3 \int \frac{\sec^2 z}{\sec^4 z} \, dz = 3 \int \tfrac{1}{2}(1 + \cos 2z) \, dz$$

$$= \tfrac{3}{2}(z + \sin z \cos z) + C_2 = \tfrac{3}{2} \left(\tan^{-1} x + \frac{x}{1 + x^2} \right) + C_2.$$

When we assemble this work, we find:

$$\int \frac{x^2 + 4}{(x^2 + 1)^2 (x^2 + 2)} \, dx = \frac{3x}{2(1 + x^2)} - \frac{1}{2} \tan^{-1} x + \sqrt{2} \tan^{-1} \left(\tfrac{1}{2}x\sqrt{2} \right) + C.$$

37. Let $x = e^{2t}$. Then $dx = 2e^{2t} \, dt$; $e^{4t} \, dt = \tfrac{1}{2}x \, dx.$

$$\int \frac{e^{4t}}{(e^{2t} - 1)^3} \, dt = \tfrac{1}{2} \int \frac{x}{(x - 1)^3} \, dx = \tfrac{1}{2} \int \left(\frac{1}{(x - 1)^2} + \frac{1}{(x - 1)^3} \right) dx$$

$$= -\tfrac{1}{4} \left(\frac{2}{x - 1} + \frac{1}{(x - 1)^2} \right) + C = -\frac{2x - 1}{4(x - 1)^2} + C = \frac{1 - 2e^{2t}}{4(e^{2t} - 1)^2} + C.$$

40. Let $u = \tan t$; $du = \sec^2 t \, dt.$ Then

$$\int \frac{\sec^2 t}{\tan^3 t + \tan^2 t} \, dt = \int \frac{1}{u^3 + u^2} \, du = \int \left(-\frac{1}{u} + \frac{1}{u^2} + \frac{1}{u + 1} \right) du$$

$$= \ln \left| \frac{u+1}{u} \right| - \frac{1}{u} + C = \ln \left| \frac{1+\tan t}{\tan t} \right| - \cot t + C$$

$$= \ln |1 + \cot t| - \cot t + C.$$

43. $\displaystyle\int_1^2 \left(\frac{\frac{5}{3}}{x} - \frac{\frac{4}{3}}{x-3} - \frac{\frac{7}{3}}{x+3} \right) dx = \frac{1}{3}(23 \ln 2 - 7 \ln 5) \approx 1.558773255.$

46. $\displaystyle\int_0^2 2\pi x \cdot \frac{x+5}{3+2x-x^2} dx = 2\pi \int_0^2 \left(\frac{6}{3-x} - 1 - \frac{1}{x+1} \right) dx$

$$= 2\pi \left[-6 \ln |3-x| - x - \ln |x+1| \right]_0^2$$

$$= 2\pi(-2 + 5 \ln 3) \approx 21.947552338.$$

49. $\displaystyle\int_1^2 \pi \left(\frac{x-9}{x^2-3x} \right)^2 dx = \pi \int_1^2 \left(\frac{4}{x} - \frac{4}{x-3} + \frac{9}{x^2} + \frac{4}{(x-3)^2} \right) dx$

$$= \pi \left[4 \ln |x| - 4 \ln |x-3| - \frac{9}{x} - \frac{4}{x-3} \right]_1^2$$

$$= \pi \left(\frac{13}{2} + 8 \ln 2 \right) \approx 37.841040971.$$

52. **(a)** The volume of revolution around the x-axis is

$$\int_0^1 \pi \cdot \frac{x^4(1-x)^2}{(1+x)^2} dx = \pi \int_0^1 \left(x^4 - 4x^3 + 8x^2 - 12x + 16 - \frac{20}{x+1} + \frac{4}{(x+1)^2} \right) dx$$

$$= \pi \left[\frac{1}{5}x^5 - x^4 + \frac{8}{3}x^3 - 6x^2 + 16x - 20 \ln |x+1| - \frac{4}{x+1} \right]_0^1$$

$$= \pi \left(\frac{208}{15} - 20 \ln 2 \right) \approx 0.011696324.$$

(b) The volume of revolution around the y-axis is

$$2 \int_0^1 2\pi x \cdot \frac{x^2(1-x)}{1+x} dx = 2\pi \int_0^1 \left(4 - 4x + 4x^2 - 2x^3 - \frac{4}{x+1} \right) dx$$

$$= 2\pi \left[4x - 2x^2 + \frac{4}{3}x^3 - \frac{1}{2}x^4 - 4 \ln |x+1| \right]_0^1$$

$$= 2\pi \left(\frac{17}{6} - 4 \ln 2 \right) \approx 0.381669648.$$

55. $\displaystyle\int \left(\frac{48}{(x-4)^2} - \frac{104}{3(x-4)} + \frac{567}{16(x-3)} - \frac{39}{2(x+5)^2} - \frac{37}{48(x+5)} \right) dx$

$$= -\frac{48}{x-4} + \frac{39}{2(x+5)} + \frac{567 \ln |x-3|}{16} - \frac{104 \ln |x-4|}{3} - \frac{37 \ln |x+5|}{48} + C.$$

58. $x(t) = \dfrac{10e^{10t}}{9 + e^{10t}}$

64. $N(t) = 10000$ when $t = \dfrac{\ln 4}{0.15} \approx 9.24$ (days).

Section 9.6

1. Let $x = 4 \sin u$. Then $16 - x^2 = 16 \cos^2 u$ and $dx = 4 \cos u \, du$.

$$\int \frac{1}{\sqrt{16 - x^2}} dx = \int \frac{4 \cos u}{4 \cos u} du = u + C = \arcsin \frac{x}{4} + C$$

4. Let $x = 5 \sec u$. Then $x^2 - 25 = 25\left(\sec^2 u - 1\right) = 25 \tan^2 u$ and $dx = 5 \sec u \tan u \, du$.

$$\int \frac{1}{x^2\sqrt{x^2 - 25}} \, dx = \int \frac{5 \sec u \tan u}{\left(25 \sec^2 u\right)\left(5 \tan u\right)} \, du = \frac{1}{25} \int \cos u \, du = \frac{1}{25} \sin u + C = \frac{\sqrt{x^2 - 25}}{25x} + C$$

7. Let $x = \frac{3}{4} \sin u$. Then $9 - 16x^2 = 9\cos^2 u$ and $dx = \frac{3}{4} \cos u \, du$.

$$\int \frac{1}{(9 - 16x^2)^{3/2}} \, dx = \int \frac{1}{27\cos^3 u}\left(\tfrac{3}{4} \cos u\right) du = \frac{1}{36} \int \sec^2 u \, du = \frac{1}{36} \tan u + C = \frac{x}{9\sqrt{9 - 16x^2}} + C$$

10. Let $x = 2 \sin u$. Then $dx = 2 \cos u \, du$ and $\sqrt{4 - x^2} = 2 \cos u$.

$$\int x^3 \sqrt{4 - x^2} \, dx = \int 32 \sin^3 u \, \cos^2 u \, du = 32 \int (\sin u)\left(\cos^2 u - \cos^4 u\right) du$$

$$= 32\left(\tfrac{1}{5} \cos^5 u - \tfrac{1}{3} \cos^3 u\right) + C = \frac{32}{15}\left(3\left(\frac{\sqrt{4 - x^2}}{2}\right)^5 - 5\left(\frac{\sqrt{4 - x^2}}{2}\right)^3\right) + C$$

$$= \tfrac{4}{15}\left(\tfrac{3}{4}\left(4 - x^2\right)\left(4 - x^2\right)^{3/2} - 5\left(4 - x^2\right)^{3/2}\right) + C$$

$$= -\tfrac{1}{15}\left(4 - x^2\right)^{3/2}\left(3x^2 + 8\right) + C$$

13. Let $x = \frac{1}{2} \sin \theta$. Then $1 - 4x^2 = \cos^2 \theta$, and the integrand becomes $\csc \theta - \sin \theta$.

16. Let $x = \frac{1}{2} \tan z$. Then $2x = \tan z$, $1 + 4x^2 = \sec^2 z$, and $dx = \frac{1}{2} \sec^2 z \, dz$.

$$\int \sqrt{1 + 4x^2} \, dx = \int \tfrac{1}{2}(\sec z)\left(\sec^2 z\right) dz = \tfrac{1}{2} \int \sec^3 z \, dz$$

Now do a moderately difficult integration by parts, or apply Formula 28 of the endpapers, to obtain

$$\int \sqrt{1 + 4x^2} \, dx = \tfrac{1}{4}\left(\sec z \tan z + \ln|\sec z + \tan z|\right) + C$$

$$= \tfrac{1}{4}\left(2x\sqrt{1 + 4x^2} + \ln\left|2x + \sqrt{1 + 4x^2}\,\right|\right) + C$$

19. Use the substitution $x = \tan u$ to transform the integrand into $\tan^2 u \sec u = \sec^3 u - \sec u$, then apply Formulas 14 and 28 of the endpapers.

22. Let $x = \sin z$. Then $\sqrt{1 - x^2} = \cos z$ and $dx = \cos z \, dz$.

$$\int \left(1 - x^2\right)^{3/2} dx = \int \cos^4 z \, dz$$

$$= \tfrac{1}{4} \int (1 + \cos 2z)^2 \, dz = \tfrac{1}{4} \int \left(1 + 2\cos 2z + \tfrac{1}{2}(1 + \cos 4z)\right) dz$$

$$= \tfrac{1}{4}\left(\tfrac{3}{2}z + \sin 2z + \tfrac{1}{8} \sin 4z\right) + C = \tfrac{3}{8}z + \tfrac{1}{2} \sin z \, \cos z + \tfrac{1}{16} \sin 2z \, \cos 2z + C$$

$$= \tfrac{3}{8}z + \tfrac{1}{2} \sin z \, \cos z + \tfrac{1}{8}(\sin z \, \cos z)\left(\cos^2 z - \sin^2 z\right) + C$$

$$= \tfrac{3}{8}z + \tfrac{1}{2} \sin z \, \cos z + \tfrac{1}{8} \sin z \, \cos^3 z - \tfrac{1}{8} \sin^3 z \, \cos z + C$$

$$= \tfrac{3}{8} \arcsin x + \tfrac{1}{2}x\sqrt{1 - x^2} + \tfrac{1}{8}x\left(1 - x^2\right)^{3/2} - \tfrac{1}{8}x^3\sqrt{1 - x^2} + C$$

$$= \tfrac{1}{8}\left(3 \arcsin x + x\left(5 - 2x^2\right)\sqrt{1 - x^2}\right) + C$$

25. The substitution $x = 2 \sin u$ transforms the integral into $\frac{1}{32} \int \sec^5 u \, du$. Application of Formula 37 of the endpapers transforms this integral into

$$\tfrac{1}{32} \int \tfrac{1}{4} \sec^3 u \, \tan u \, du + \tfrac{3}{4} \int \sec^3 u \, du.$$

Then Formula 28 from the endpapers yields the antiderivative

$$\tfrac{1}{128}\sec^3 u\,\tan u + \tfrac{3}{256}\sec u\,\tan u + \tfrac{3}{256}\ln|\sec u + \tan u| + C.$$

Finally make the replacements $\sec u = \dfrac{2}{\sqrt{4-x^2}}$ and $\tan u = \dfrac{x}{\sqrt{4-x^2}}$ to obtain the final answer.

28. Use the substitution $x = \tfrac{3}{4}\tan u$ to obtain $\tfrac{81}{4}\displaystyle\int \sec^5 u\,du$. With the aid of Formula 37 of the endpapers, we next obtain

$$\int \left(9 + 16x^2\right)^{3/2}\,dx = \tfrac{81}{4}\left(\tfrac{1}{4}\sec^3 u\,\tan u + \tfrac{3}{8}\sec u\,\tan u + \tfrac{3}{8}\ln|\sec u + \tan u|\right) + C$$

$$= \tfrac{1}{4}x\left(9 + 16x^2\right)^{3/2} + \tfrac{27}{8}x\left(9 + 16x^2\right)^{1/2} + \tfrac{243}{32}\ln\left|4x + \sqrt{9 + 16x^2}\,\right| + C.$$

(We have allowed the constant $-\tfrac{243}{32}\ln 3$ to be "absorbed" by the constant C.)

31. Let $x = \sec u$. Then $x^2 - 1 = \tan^2 x$ and $dx = \sec u\,\tan u\,du$.

$$\int x^2\sqrt{x^2 - 1}\,dx = \int \sec^3 u\,\tan^2 u\,du = \int \left(\sec^5 u - \sec^3 u\right)\,du$$

$$= \tfrac{1}{4}\sec^3 u\,\tan u - \tfrac{1}{8}\sec u\,\tan u - \tfrac{1}{8}\ln|\sec u + \tan u| + C$$

—with the aid of Formula 37 of the endpapers. Then replace $\tan u$ with $\sqrt{x^2 - 1}$ and $\sec u$ with x to obtain the final version of the answer.

34. Let $x = \tfrac{3}{2}\sec u$. Then $\sqrt{4x^2 - 9} = \sqrt{9\sec^2 u - 9} = 3\tan u$ and $dx = \tfrac{3}{2}\sec u\,\tan u\,du$.

$$\int \frac{1}{x^2\sqrt{4x^2 - 9}}\,dx = \tfrac{2}{9}\int \cos u\,du = \tfrac{2}{9}\sin u + C = \frac{\sqrt{4x^2 - 9}}{9x} + C$$

37. The substitution $x = 5\sinh u$ yields

$$\int \frac{dx}{\sqrt{25 + x^2}}\,dx = \int \frac{5\cosh u}{5\cosh u}\,du = \int 1\,du = u + C = \sinh^{-1}\left(\frac{x}{5}\right) + C.$$

40. Let $x = \tfrac{1}{3}\sinh u$. Then $9x^2 = \sinh^2 u$, $\sqrt{1 + 9x^2} = \cosh u$, and $dx = \tfrac{1}{3}\cosh u\,du$.

$$\int \frac{dx}{\sqrt{1 + 9x^2}}\,dx = \tfrac{1}{3}\int \frac{\cosh u}{\cosh u}\,du = \tfrac{1}{3}u + C = \tfrac{1}{3}\sinh^{-1}(3x) + C$$

46. The length of one arch of the sine curve is

$$S = \int_0^\pi \sqrt{1 + \cos^2 x}\,dx.$$

To obtain the length of the upper half of the ellipse, take

$$y = \sqrt{2 - 2x^2}, \qquad -1 \le x \le 1.$$

Then $\dfrac{dy}{dx} = -\dfrac{2x}{\sqrt{2 - 2x^2}}$, so—after algebraic simplification—the arc length is

$$E = \int_{-1}^1 \frac{\sqrt{1 + x^2}}{\sqrt{1 - x^2}}\,dx.$$

Let $x = \cos u$. Then

$$E = \int_\pi^0 \frac{\sqrt{1 + \cos^2 u}}{\sqrt{1 - \cos^2 u}}(-\sin u)\,du = \int_0^\pi \sqrt{1 + \cos^2 u}\,du = S.$$

49. After algebraic simplification, the surface area integral takes the form

$$A = 4\pi \int_{a-b}^{a+b} \frac{bx}{\sqrt{b^2 - (x-a)^2}} \, dx.$$

The substitution $x = a + b\sin\theta$ leads to the desired result.

52. After it is simplified, the surface area integral should become

$$A = \frac{4\pi b}{a^2} \int_0^a \sqrt{a^4 - x^2 \left(a^2 - b^2\right)} \, dx.$$

To find the antiderivative, let $x = \dfrac{a^2 \sin u}{\sqrt{a^2 - b^2}}$. Note that $\arcsin w \approx w$ if $w \approx 0$.

55. One integral that yields the cost (in millions of dollars) is

$$C = \int_2^5 \frac{x}{\sqrt{x-1}} \, dx.$$

The substitution $x = \sec^2\theta$ yields the antiderivative

$$\tfrac{2}{3} \sec^2\theta \tan\theta + \tfrac{4}{3} \tan\theta + C = \tfrac{2}{3} x\sqrt{x-1} + \tfrac{4}{3}\sqrt{x-1} + C.$$

It follows that the total cost of the road will be $\frac{20}{3}$ million dollars.

Section 9.7

1. $\displaystyle\int \frac{1}{x^2 + 4x + 5} \, dx = \int \frac{1}{(x+2)^2 + 1} \, dx = \arctan(x+2) + C.$

4. Let $u = x + 2$: $x = u - 2$ and $dx = du$. Thus

$$\int \frac{x+1}{(x^2 + 4x + 5)^2} \, dx = \int \frac{u-1}{(u^2 + 1)^2} \, du.$$

Next let $u = \tan\theta$. Then $du = \sec^2\theta \, d\theta$, and

$$\int \frac{x+1}{(x^2 + 4x + 5)^2} \, dx = \int \frac{\tan\theta - 1}{\sec^2\theta} \, d\theta = \int \left(\sin\theta \cos\theta - \cos^2\theta\right) d\theta$$

$$= \tfrac{1}{2}\sin^2\theta - \tfrac{1}{2}\theta - \tfrac{1}{2}\sin\theta \cos\theta + C$$

$$= \frac{u^2}{2(1+u^2)} - \tfrac{1}{2}\arctan u - \frac{u}{2(1+u^2)} + C$$

$$= \frac{x^2 + 3x + 2}{2(x^2 + 4x + 5)} - \tfrac{1}{2}\arctan(x+2) + C.$$

7. First, $3 - 2x - x^2 = 4 - (x+1)^2$, so we let $x = -1 + 2\sin u$: $x + 1 = 2\sin u$, $dx = 2\cos u \, du$. Then $4 - (x+1)^2 = 4 - 4\sin^2 u = 4\cos^2 u$, so

$$\int x\sqrt{3 - 2x - x^2} \, dx = \int (-1 + 2\sin u)(2\cos u)(2\cos u) \, du$$

$$= \int \left(-4\cos^2 u + 8\cos^2 u \sin u\right) du$$

$$= \int \left(-2 - 2\cos 2u + 8\cos^2 u \sin u\right) du$$

$$= -2u - 2\sin u \cos u - \tfrac{8}{3}\cos^3 u + C$$

$$= -2\arcsin\left(\frac{x+1}{2}\right) - \tfrac{1}{2}(x+1)\sqrt{3 - 2x - x^2} - \tfrac{1}{3}\left(3 - 2x - x^2\right)^{3/2} + C.$$

10. Because $4x^2 + 4x - 3 = (2x+1)^2 - 4$, we let $x = -\frac{1}{2} + \sec u$: $dx = \sec u \tan u\, du$, $2x + 1 = 2\sec u$, and $(2x+1)^2 - 4 = 4\tan^2 u$.

$$\int \sqrt{4x^2 + 4x - 3}\, dx = \int 2\sec u \tan^2 u\, du = 2\int (\sec^3 u - \sec u)\, du$$
$$= \sec u \tan u - \ln|\sec u + \tan u| + C$$
$$= \tfrac{1}{4}(2x+1)(4x^2 + 4x - 3)^{1/2} - \ln\left|2x+1+\sqrt{4x^2+4x-3}\right| + C$$

(the constant $\ln 2$ has been "absorbed" by C).

13. $3 + 2x - x^2 = 4 - (x-1)^2 = 4 - 4\sin^2 u$ if we let $x = 1 + 2\sin u$. Then $dx = 2\cos u\, du$, and therefore

$$\int \frac{1}{3+2x-x^2}\, dx = \int \frac{2\cos u}{4\cos^2 u}\, du = \tfrac{1}{2}\int \sec u\, du$$
$$= \tfrac{1}{2}\ln|\sec u + \tan u| + C = \tfrac{1}{2}\ln\left|\frac{x+1}{\sqrt{3+2x-x^2}}\right| + C.$$

16. Since $4x^2 + 4x - 15 = (2x+1)^2 - 16$, we let $2x + 1 = 4\sec z$. Then $x = \tfrac{1}{2}(-1 + 4\sec z)$ and $dx = 2\sec z \tan z\, dz$.

$$\int \frac{2x-1}{4x^2+4x-15}\, dx = \int \frac{4\sec z - 2}{16\tan^2 z}(2\sec z \tan z)\, dz$$
$$= \tfrac{1}{4}\int \frac{(2\sec z - 1)\sec z}{\tan z}\, dz$$
$$= \tfrac{1}{2}\int \frac{\sec^2 z}{\tan z}\, dz - \tfrac{1}{4}\int \csc z\, dz$$
$$= \tfrac{1}{2}\ln|\tan z| - \tfrac{1}{4}\ln|\csc z - \cot z| + C_1.$$

With the aid of the reference triangle to the right, we can "translate" the expression above into a function of the original variable x. Thus we find the antiderivative to be

$$\tfrac{1}{2}\ln\frac{\sqrt{4x^2+4x-15}}{4} - \tfrac{1}{4}\ln\frac{2x+1-4}{\sqrt{4x^2+4x-15}} + C_1,$$

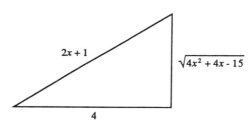

which can be simplified to

$$\tfrac{1}{8}\ln|2x-3| - \tfrac{3}{8}\ln|2x+5| + C.$$

19. First, $9 + 16x - 4x^2 = 25 - (2x-4)^2$. Then use the trigonometric substitution $x = 2 + \frac{5}{2}\sin u$.

$$\int (7-2x)\sqrt{9+16x-4x^2}\, dx = \int \left(\tfrac{75}{2}\cos^2 u - \tfrac{125}{2}\cos^2 u \sin u\right)\, du$$
$$= \tfrac{25}{12}(9u + 9\sin u \cos u + 10\cos^3 u) + C.$$

22. Let $x = \tan u$. Then

$$\int \frac{x-1}{(x^2+1)^2}\, dx = \int \frac{\tan u - 1}{\sec^4 u}\sec^2 u\, du = \int \left(\frac{\tan u}{\sec^2 u} - \frac{1}{\sec^2 u}\right)\, du = \int (\cos u \sin u - \cos^2 u)\, du$$
$$= \tfrac{1}{2}\sin^2 u - \tfrac{1}{2}u - \tfrac{1}{2}\sin u \cos u + C$$
$$= \frac{x^2}{2(x^2+1)} - \tfrac{1}{2}\tan^{-1} x - \frac{x}{2(x^2+1)} + C$$
$$= \frac{x(x-1)}{2(x^2+1)} - \tfrac{1}{2}\tan^{-1} x + C.$$

25. We note that $x^2 + x + 1 = \frac{3}{4}\left(\frac{4}{3}\left(x + \frac{1}{2}\right)^2 + 1\right)$. This suggests that we attempt to obtain

$$\tfrac{4}{3}\left(x + \tfrac{1}{2}\right)^2 = \tan^2 u,$$

which we accomplish by using the substitution

$$x = \tfrac{1}{2}\left(-1 + \sqrt{3}\tan u\right).$$

This leads to

$$\int \frac{3x - 1}{x^2 + x + 1}\,dx = \int \left(3\tan u - \tfrac{5}{3}\sqrt{3}\right)\,du,$$

and only the final answer is complicated.

28. Because $x - x^2 = \frac{1}{4}\left(1 - (2x - 1)^2\right)$, we want $2x - 1 = \sin u$: Let $x = \frac{1}{2}(1 + \sin u)$. This yields

$$\int (x - x^2)^{2/3}\,dx = \tfrac{1}{16}\int \cos^4 u\,du.$$

Then we apply Formula 34 of the endpapers to obtain

$$\int (x - x^2)^{3/2}\,dx = \tfrac{1}{128}\left(3u + 4\sin u\,\cos u + \sin u\,\cos^3 u - \sin^3 u\,\cos u\right) + C$$

$$= \tfrac{3}{128}\sin^{-1}(2x - 1) - \tfrac{1}{64}\left(16x^3 - 24x^2 + 2x + 3\right)\sqrt{x - x^2} + C.$$

31. $\dfrac{2x^2 + 3}{x^4 - 2x^2 + 1} = \dfrac{A}{x - 1} + \dfrac{B}{(x - 1)^2} + \dfrac{C}{x + 1} + \dfrac{D}{(x + 1)^2}$ leads to $A = -\frac{1}{4}$, $B = \frac{5}{4}$, $C = \frac{1}{4}$, and $D = \frac{5}{4}$, and thus:

$$\int \frac{2x^2 + 3}{x^4 - 2x^2 + 1} = \tfrac{1}{4}\ln\left|\frac{x + 1}{x - 1}\right| - \frac{5x}{2\left(x^2 - 1\right)} + C.$$

34. $\displaystyle \int \frac{x^3 - 2x}{x^2 + 2x + 2}\,dx = \int x - 2 + \frac{4}{1 + (x + 1)^2}\,dx = \tfrac{1}{2}x^2 - 2x + 4\arctan(x + 1) + C.$

Originally, Problem 34 was supposed to read

$$\int \frac{x^3 - 2x}{(x^2 + 2x + 2)^2}\,dx,$$

which is somewhat more challenging. The antiderivative in this case is

$$\tfrac{1}{2}\ln(x^2 + 2x + 2) - \arctan(x + 1) + \frac{2(x + 1)}{x^2 + 2x + 2} + C.$$

37. The area of R is $\displaystyle \int_0^5 \frac{1}{x^2 - 2x + 5}\,dx = \left[\tfrac{1}{2}\arctan\left(\tfrac{1}{2}[x - 1]\right)\right]_0^5 = \tfrac{1}{4}\pi.$

40. The area of R is

$$\int_1^4 \frac{1}{4x^2 - 20x + 29}\,dx = \left[\tfrac{1}{4}\arctan\left(\tfrac{1}{2}[2x - 5]\right)\right]_1^4$$

$$= \tfrac{1}{2}\arctan\left(\tfrac{3}{2}\right) \approx 0.491396862.$$

43. First, $y = \frac{19}{4} + \sqrt{R^2 - (x + 1)^2}$ where $R = \sqrt{377/16}$. It follows that $\dfrac{dy}{dx} = -\dfrac{x + 1}{\sqrt{R^2 - (1 + x)^2}}$ and

hence the length of the road is given by

$$S = \int_0^3 R\left(R^2 - (1 + x)^2\right)^{-1/2}\,dx$$

$$= R\left[\sin^{-1}\left(\frac{1 + x}{R}\right)\right]_0^3 = -\tfrac{1}{4}\sqrt{377}\arctan\tfrac{20}{21} \approx 3.6940487.$$

Thus the length of the road is just over 3.69 miles.

46. $\dfrac{1}{x^3 + 8} = \dfrac{A}{x + 2} + \dfrac{Bx + C}{x^2 - 2x + 4}$ leads to $A = \frac{1}{12}$, $B = -\frac{1}{12}$, and $C = \frac{1}{3}$. Thus

$$\int \frac{1}{x^3 + 8}\, dx = \frac{1}{12} \int \left(\frac{1}{x + 2} + \frac{4 - x}{x^2 - 2x + 4} \right) dx$$

$$= \frac{1}{12} \int \left(\frac{1}{x + 2} - \frac{1}{2} \left(\frac{2x - 2}{x^2 - 2x + 4} \right) + \frac{3}{x^2 - 2x + 4} \right) dx.$$

Note that $x^2 - 2x + 4 = (x - 1)^2 + 3$, and with the aid of Formula 17 of the endpapers we obtain the final answer; after simplification, it is

$$\int \frac{1}{x^3 + 8}\, dx = \frac{1}{24} \ln \left| \frac{x^2 + 4x + 4}{x^2 - 2x + 4} \right| + \frac{1}{12} \sqrt{3} \arctan \left(\tfrac{1}{3}\sqrt{3}\,(x - 1) \right) + C.$$

49. $x^4 + x^2 + 1 = (x^2 + x + 1)(x^2 - x + 1)$. (It follows that $10^{12} + 10^6 + 1 = 1,000,001,000,001$ is composite.) And

$$\frac{2x^3 + 3x}{x^4 + x^2 + 1} = \frac{Ax + B}{x^2 + x + 1} + \frac{Cx + D}{x^2 - x + 1}$$

yields the values $A = 1$, $B = -\frac{1}{2}$, $C = 1$, and $D = \frac{1}{2}$. So

$$\int \frac{2x^3 + 3x}{x^4 + x^2 + 1}\, dx = \frac{1}{2} \int \left(\frac{2x - 1}{x^2 + x + 1} + \frac{2x + 1}{x^2 - x + 1} \right) dx$$

$$= \frac{1}{2} \int \left(\frac{2x + 1}{x^2 + x + 1} - \frac{2}{x^2 + x + 1} + \frac{2x - 1}{x^2 - x + 1} + \frac{2}{x^2 - x + 1} \right) dx$$

$$= \frac{1}{2} \left(\ln (x^2 + x + 1) + \ln (x^2 - x + 1) \right) + \int \left(\frac{1}{\left(x - \frac{1}{2}\right)^2 + \frac{3}{4}} - \frac{1}{\left(x + \frac{1}{2}\right)^2 + \frac{3}{4}} \right) dx$$

$$= \frac{1}{2} \ln (x^4 + x^2 + 1) + \frac{2}{3}\sqrt{3} \int \left(\frac{\frac{2}{3}\sqrt{3}}{\left(\frac{2x - 1}{\sqrt{3}}\right)^2 + 1} - \frac{\frac{2}{3}\sqrt{3}}{\left(\frac{2x + 1}{\sqrt{3}}\right)^2 + 1} \right) dx$$

$$= \frac{1}{2} \ln (x^4 + x^2 + 1) + \frac{2}{3}\sqrt{3} \left(\arctan \left(\frac{2x - 1}{\sqrt{3}} \right) - \arctan \left(\frac{2x + 1}{\sqrt{3}} \right) \right) + C$$

$$= \frac{1}{2} \ln (x^4 + x^2 + 1) + \frac{2\sqrt{3}}{3} \arctan \left(\frac{\sqrt{3}}{3}\,(2x^2 + 1) \right) + C.$$

52. The partial fraction decomposition of the integrand is $\dfrac{7x + 3}{(x^2 + 1)^2} - \dfrac{4x - 5}{x^2 + 6x + 10}$.

The antiderivative is $\dfrac{3x - 7}{2(x^2 + 1)} + \frac{3}{2} \arctan x + 17 \arctan(x + 3) - 2 \ln(x^2 + 6x + 10) + C$.

Section 9.8

1. $\displaystyle \int_4^\infty x^{-3/2}\, dx = \lim_{t \to \infty} \left[\frac{2}{x^{1/2}} \right]_t^4 = 1.$

4. $\displaystyle \int_0^8 x^{-2/3}\, dx = \lim_{t \to 0} \left[3x^{1/3} \right]_t^8 = 6.$

7. $\displaystyle \int_5^\infty (x - 1)^{-3/2}\, dx = \lim_{t \to \infty} \left[-\frac{2}{\sqrt{x - 1}} \right]_5^t = 1.$

10. $\displaystyle\int_0^3 \frac{1}{(x-3)^2}\, dx = \lim_{t\to 3^-}\left[-\frac{1}{x-3}\right]_0^t$ diverges.

13. Integrate from -1 to 0, from 0 to 8. If both integrals converge add the results.

16. $\displaystyle\int_{-\infty}^\infty \frac{x}{(x^2+4)^{3/2}}\, dx = \int_{-\infty}^0 \frac{x}{(x^2+4)^{3/2}}\, dx + \int_0^\infty \frac{x}{(x^2+4)^{3/2}}\, dx$

$$= \lim_{s\to-\infty}\left[-\frac{1}{\sqrt{x^2+4}}\right]_s^0 + \lim_{t\to\infty}\left[-\frac{1}{\sqrt{x^2+4}}\right]_0^t = 0.$$

22. $\displaystyle\left[\tfrac{1}{2}e^{2x}\right]_{-\infty}^2 = \tfrac{1}{2}e^4.$

25. $\displaystyle\left[\arctan x\right]_0^\infty = \frac{\pi}{2}.$

28. This integral diverges to $+\infty$.

31. $\displaystyle\left[-\frac{1}{\ln x}\right]_2^\infty = \frac{1}{\ln 2}.$

34. This integral diverges to $+\infty$.

37. This integral diverges to $-\infty$.

40. $\displaystyle\int_0^1 \frac{1}{x^2+x^4}\, dx = +\infty;\ \int_1^\infty \left(\frac{1}{x^2}-\frac{1}{x^2+1}\right)dx = \left[-\frac{1}{x}-\arctan x\right]_1^\infty = \frac{4-\pi}{4}.$

43. This integral converges exactly when $k < 1$. For such k, its value is $\displaystyle\left[\frac{x^{1-k}}{1-k}\right]_0^1 = \frac{1}{1-k}.$

46. Note that $\displaystyle\int \frac{1}{x(\ln x)^k}\, dx = \frac{(\ln x)^{1-k}}{1-k} + C.$ If $1-k > 0$, then the improper integral will diverge because $(\ln x)^{1-k} \to +\infty$ as $x \to +\infty$. If $1-k < 0$, then the improper integral will diverge because $(\ln x)^{1-k} = \dfrac{1}{(\ln x)^{k-1}} \to +\infty$ as $x \to 1^+$. The integral clearly diverges if $k = 1$. Therefore this integral converges for *no* values of k.

52. $\displaystyle\int_0^\infty \frac{1+x}{1+x^2}\, dx$ dominates $\displaystyle\int_0^\infty \frac{x}{1+x^2}\, dx$, which diverges.

58. We need to evaluate $V = \displaystyle\int_0^\infty \pi[\exp(-x^2)]^2\, dx = \int_0^\infty \pi\exp(-2x^2)\, dx.$

Let $x = u/\sqrt{2}$. Then $V = \displaystyle\int_{u=0}^\infty \frac{1}{\sqrt{2}}\,\pi e^{-u^2}\, du = \frac{\pi}{\sqrt{2}}\int_0^\infty e^{-u^2}\, du = \frac{\pi}{\sqrt{2}}\cdot\frac{\sqrt{\pi}}{2} = \left(\frac{\pi}{2}\right)^{3/2}.$

61. **(a)** $\displaystyle\left[\begin{array}{ll} u = x^{k-1} & dv = xe^{-x^2}\, dx \\ du = (k-1)\, x^{k-2}\, dx & v = -\tfrac{1}{2}e^{-x^2} \end{array}\right]$

$$\int_0^\infty x^k\, e^{-x^2}\, dx = \left[-\frac{x^{k-1}}{2}\, e^{-x^2}\right]_0^\infty + \frac{k-1}{2}\int_0^\infty x^{k-2}\, e^{-x^2}\, dx = \frac{k-1}{2}\int_0^\infty x^{k-2}\, e^{-x^2}\, dx.$$

(b) If $n = 1$, then

$$\int_0^\infty x^{n-1}\, e^{-x^2}\, dx = \int_0^\infty e^{-x^2}\, dx = \tfrac{1}{2}\sqrt{\pi}\ \text{(by Problem 35)}$$
$$= \tfrac{1}{2}\Gamma\left(\tfrac{1}{2}\right)\ \text{(also by Problem 35).}$$

If $n = 2$, then

$$\int_0^\infty x^{n-1}\, e^{-x^2}\, dx = \int_0^\infty xe^{-x^2}\, dx = \left[-\tfrac{1}{2}e^{-x^2}\right]_0^\infty = \tfrac{1}{2} = \tfrac{1}{2}\Gamma(1)\ \text{because } \Gamma(1) = 0! = 1.$$

If $n = 3$, then

$$\int_0^\infty x^{n-1} e^{-x^2} \, dx = \frac{n-2}{2} \int_0^\infty x^{n-3} e^{-x^2} \, dx \quad \text{(by Part (a))}$$

$$= \tfrac{1}{2} \int_0^\infty e^{-x^2} \, dx = \tfrac{1}{2} \cdot \tfrac{1}{2}\sqrt{\pi}$$

$$= \tfrac{1}{2}\Gamma\left(\tfrac{3}{2}\right) \text{ because } \Gamma\left(\tfrac{3}{2}\right) = \tfrac{1}{2}\Gamma\left(\tfrac{1}{2}\right) = \tfrac{1}{2}\sqrt{\pi}.$$

Assume that for some integer $k \geq 3$,

$$\int_0^\infty x^{n-1} e^{-x^2} \, dx = \tfrac{1}{2}\Gamma\left(\frac{n}{2}\right) \text{ for all } n, \ 1 \leq n \leq k.$$

Then

$$\int_0^\infty x^k e^{-x^2} \, dx = \frac{k-1}{2} \int_0^\infty x^{k-2} e^{-x^2} \, dx = \frac{k-1}{2} \cdot \tfrac{1}{2}\Gamma\left(\frac{k-1}{2}\right) = \tfrac{1}{2}\Gamma\left(\frac{k+2}{2}\right)$$

(because $\Gamma(x+1) = x\,\Gamma(x)$). Therefore, by induction, $\int_0^\infty x^{n-1} e^{-x^2} \, dx = \tfrac{1}{2}\Gamma\left(\frac{n}{2}\right)$ for all $n \geq 1$.

Chapter 9 Miscellaneous

4. $\displaystyle\int \frac{\csc x \cot x}{1 + \csc^2 x} \, dx = -\tan^{-1}(\csc x) + C$

7. Integration by parts: Let $u = x$, $dv = \tan^2 x \, dx = (\sec^2 x - 1) \, dx$.

10. Let $x = 2 \tan u$:

$$\int \frac{1}{\sqrt{x^2 + 4}} \, dx = \int \sec u \, du = \ln|\sec u + \tan u| + C$$

$$= \ln\left|\tfrac{1}{2}\sqrt{x^2 + 4} + \tfrac{1}{2}x\right| + C_1 = \ln\left|x + \sqrt{x^2 + 4}\right| + C.$$

13. Write $x^2 - x + 1 = \left(x - \tfrac{1}{2}\right)^2 + \tfrac{3}{4}$, then apply Formula 17 from the endpapers.

16. $\displaystyle\int \frac{x^4 + 1}{x^2 + 2} \, dx = \int \left(x^2 - 2 + \frac{5}{x^2 + 2}\right) dx = \tfrac{1}{3}x^3 - 2x + \tfrac{5}{2}\sqrt{2} \tan^{-1}\left(\tfrac{1}{2}x\sqrt{2}\right) + C.$

19. Use the substitution $u = \sin x$.

22. Let $x^2 = \sin u$:

$$\int \frac{x^7}{\sqrt{1 - x^4}} \, dx = \int \tfrac{1}{2}\sin^3 u \, du = \tfrac{1}{6}(\cos u)(\cos^2 u - 3) + C = -\tfrac{1}{6}(x^4 + 2)\sqrt{1 - x^4} + C.$$

25. Let $x = 3 \tan u$.

28. $\displaystyle\int \frac{4x - 2}{x^3 - x} \, dx = \int \left(\frac{2}{x} - \frac{3}{x + 1} + \frac{1}{x - 1}\right) dx$

$$= 2 \ln|x| - 3 \ln|x + 1| + \ln|x - 1| + C = \ln\left|\frac{x^2(x - 1)}{(x + 1)^3}\right| + C.$$

31. Let $x = -1 + \tan u$: $\displaystyle\int \frac{x}{(x^2 + 2x + 2)^2} \, dx = \int \frac{-1 + \tan u}{\sec^2 u} \, du = \int (-\cos^2 u + \sin u \cos u) \, du$, and the rest is routine.

34. $\displaystyle\int \frac{\sec x}{\tan x} \, dx = \int \csc x \, dx = \ln|\csc x - \cot x| + C.$

37. *Suggestion:* Develop a reduction formula for $\int x (\ln x)^n\, dx$ by parts; take $u = (\ln x)^n$ and $dv = x\, dx$. Then apply your formula iteratively to evaluate the given antiderivative. You should find that

$$\int x (\ln x)^n\, dx = \tfrac{1}{2}x^2 (\ln x)^n - \frac{n}{2}\int x (\ln x)^{n-1}\, dx.$$

40. Note that $4x - x^2 = 4 - (x-2)^2$; let $x - 2 = 2\sin u$. Then

$$\int \frac{x}{\sqrt{4x - x^2}}\, dx = \int 2(1 + \sin u)\, du = 2(u - \cos u) + C = 2\arcsin\frac{x-2}{2} - \sqrt{4x - x^2} + C.$$

43. Use the method of partial fraction decomposition.

46. Here is a quick way to obtain the partial fraction decomposition: $\dfrac{x^2 + 2x + 2}{(x+1)^3} = \dfrac{(x+1)^2 + 1}{(x+1)^3}$.

$$\int \frac{x^2 + 2x + 2}{(x+1)^3}\, dx = \int \left(\frac{1}{x+1} + \frac{1}{(x+1)^3}\right) dx = \ln|x+1| - \frac{1}{2(x+1)^2} + C.$$

49. The partial fraction decomposition of the integrand has the form

$$\frac{A}{x-1} + \frac{Bx + C}{x^2 + x + 1} + \frac{D}{(x-1)^2} + \frac{Ex + F}{(x^2 + x + 1)^2}.$$

The simultaneous equations are

$$
\begin{array}{rcl}
A + B & = & 3 \\
A - B + C + D & = & -1 \\
A \quad\; - C + 2D + E & = & 2 \\
-A - B \quad\; + 3D - 2E + F & = & -12 \\
-A + B - C + 2D + E - 2F & = & -2 \\
-A \quad\; + C + D \quad\; + F & = & 1
\end{array}
$$

Their solution: $A = 1$, $B = 2$, $C = 1$, $D = -1$, $E = 4$, and $F = 2$. None of the antiderivatives is difficult.

$$\int \frac{3x^5 - x^4 + 2x^3 - 12x^2 - 2x + 1}{(x^3 - 1)^2}\, dx = \ln|x-1| + \ln(x^2 + x + 1) + \frac{1}{x-1} - \frac{2}{x^2 + x + 1} + C.$$

52. Let $x = u^3$: $\displaystyle\int \frac{(1 + x^{2/3})^{3/2}}{x^{1/3}}\, dx = 3\int u(1 + u^2)^{3/2}\, du = \tfrac{3}{5}(1 + u^2)^{5/2} + C = \tfrac{3}{5}(1 + x^{2/3})^{5/2} + C.$

55. $\tan^3 z = (\tan z)(\sec^2 z - 1) = (\tan z)(\sec^2 z) - \tan z.$

58. Note that $\dfrac{\cos^3 x}{\sqrt{\sin x}} = \dfrac{(1 - \sin^2 x)(\cos x)}{\sqrt{\sin x}} = (\sin x)^{-1/2}\cos x - (\sin x)^{3/2}\cos x.$ So

$$\int \frac{\cos^3 x}{\sqrt{\sin x}}\, dx = \tfrac{2}{5}(\sin x)^{1/2}(5 - \sin^2 x) + C.$$

61. Use integration by parts with $u = \arcsin x$ and $dv = \dfrac{1}{x^2}\, dx$. Then apply Formula 60 of the endpapers, or use the trigonometric substitution $x = \sin u$.

64. Because $2x - x^2 = 1 - (x-1)^2$, let $x = 1 + \sin u$:

$$\int x\sqrt{2x - x^2}\, dx = \int (1 + \sin u)(\cos^2 u)\, du = \int \left(\frac{1 + \cos 2u}{2} + \cos^2 u \sin u \right) du$$

$$= \tfrac{1}{2}u + \tfrac{1}{2}\sin u \cos u - \tfrac{1}{3}\cos^3 u + C$$

$$= \tfrac{1}{2}\sin^{-1}(x-1) + \tfrac{1}{2}(x-1)\sqrt{2x - x^2} - \tfrac{1}{3}\left(2x - x^2\right)^{3/2} + C$$

$$= \tfrac{1}{2}\sin^{-1}(x-1) + \tfrac{1}{6}\sqrt{2x - x^2}\,(2x^2 - x - 3) + C.$$

70. Let $u = \tan x$: $\displaystyle \int \frac{\sec^2 x}{\tan^2 x + 2\tan x + 2}\, dx = \int \frac{1}{u^2 + 2u + 2}\, du = \int \frac{1}{1 + (u+1)^2}\, du$

$$= \tan^{-1}(u+1) + C = \tan^{-1}(1 + \tan x) + C.$$

73. Let $u = x^3 - 1$; the rest is routine.

76. Let $x = \tan^3 z$. $\displaystyle \int \frac{1}{x^{2/3}\left(1 + x^{2/3}\right)}\, dx = \int \frac{3\tan^2 z \sec^2 z}{\sec^2 z \tan^2 z}\, dz = 3z + C = 3\tan^{-1}\left(x^{1/3}\right) + C.$

79. Multiply numerator and denominator of the integrand by $\sqrt{1 - \sin t}$.

82. Let $u = e^x$: $\displaystyle \int e^x \sin^{-1}(e^x)\, dx = \int \sin^{-1} u\, du$. Now do an integration by parts with $p = \sin^{-1} u$, $dq = du$:

$$\int e^x \sin^{-1}(e^x)\, dx = u\sin^{-1} u - \int u\left(1 - u^2\right)^{-1/2} du$$

$$= u\sin^{-1} u + \sqrt{1 - u^2} + C = e^x \sin^{-1}(e^x) + \sqrt{1 - e^{2x}} + C.$$

85. The partial fraction decomposition of the integrand is $\dfrac{x}{x^2 + 1} - \dfrac{x}{\left(x^2 + 1\right)^2}$, and integration of these terms presents no difficulties.

88. $\begin{bmatrix} u = \ln x & dv = x^{3/2}\, dx \\ du = \dfrac{1}{x}\, dx & v = \tfrac{2}{5}x^{5/2} \end{bmatrix}$ $\displaystyle \int x^{3/2} \ln x\, dx = \tfrac{2}{25}x^{5/2}(-2 + 5\ln x) + C.$

91. Use integration by parts. A good choice is $u = x$, $dv = e^x \sin x\, dx$. One must then antidifferentiate both $e^x \sin x$ and $e^x \cos x$, but here Formulas 67 and 68 of the endpapers may be used, or integration by parts will suffice for each.

94. $\begin{bmatrix} u = \ln\left(1 + \sqrt{x}\right) & dv = dx \\ du = \dfrac{1}{2\sqrt{x}\left(1 + \sqrt{x}\right)}\, dx & v = x - 1 \end{bmatrix}$ $\displaystyle \int \ln\left(1 + \sqrt{x}\right) dx = (x-1)\ln\left(1 + \sqrt{x}\right) - \tfrac{1}{2}x + \sqrt{x} + C.$

97. Let $u = x - 1$: $\displaystyle \int \frac{x^4}{(x-1)^2}\, dx = \int \frac{(u+1)^4}{u^2}\, du = \int \left(u^2 + 4u + 6 + \frac{4}{u} + \frac{1}{u^2}\right) du.$

100. Let $u = x^2$: $\displaystyle \int x\sqrt{\frac{1 - x^2}{1 + x^2}}\, dx = \tfrac{1}{2}\int \sqrt{\frac{1 - u}{1 + u}}\, du.$

Now let $v^2 = \dfrac{1 - u}{1 + u}$; $u = \dfrac{1 - v^2}{1 + v^2}$ and $du = -\dfrac{4v}{\left(1 + v^2\right)^2}\, dv.$

$$\int x\sqrt{\frac{1 - x^2}{1 + x^2}}\, dx = -\tfrac{1}{2}\int \frac{4v^2}{\left(1 + v^2\right)^2}\, dv.$$

Finally let $v = \tan z$.

$$\int x\sqrt{\frac{1 - x^2}{1 + x^2}}\, dx = -2\int \sin^2 z\, dz = \sin z \cos z - z + C = \tfrac{1}{2}\sqrt{1 - x^4} - \tan^{-1}\frac{1 - x^2}{\sqrt{1 + x^2}} + C.$$

103. $A_t = \int_0^t 2\pi e^{-x}\sqrt{1 + e^{-2x}}\, dx$. Let $u = e^{-x}$. Then

$$A_t = \int_1^{e^{-t}} 2\pi u\sqrt{1 + u^2}\left(-\frac{1}{u}\right) du = \int_p^1 2\pi\sqrt{1 + u^2}\, du \quad \text{(where } p = e^{-t}\text{).}$$

Let $u = \tan z$. Then

$$A_t = \int_{u=p}^{u=1} 2\pi\sqrt{1 + \tan^2 z}\, \sec^2 z\, dz$$

$$= \pi\left[\sec z\,\tan z + \ln|\sec z + \tan z|\right]_{u=p}^{u=1}$$

$$= \pi\left(\sqrt{2} + \ln\left(1 + \sqrt{2}\right) - e^{-t}\sqrt{1 + e^{-2t}} - \ln\left|e^{-t} + \sqrt{1 + e^{-2t}}\right|\right).$$

$$\lim_{t\to\infty} A_t = \pi\left(\sqrt{2} + \ln\left(1 + \sqrt{2}\right)\right) \approx 7.2118.$$

106. **(a)** Use integration by parts with $u = (\ln x)^n$ and $dv = x^m\, dx$.

(b) Application of the formula and simplification of the result yields $\dfrac{17e^4 + 3}{128} \approx 7.2747543$.

109. The area is $A = 2\displaystyle\int_0^2 x^{5/2}(2 - x)^{1/2}\, dx$. Use the suggested substitution $x = 2\sin^2\theta$. This results in $A = 64\displaystyle\int_0^{\pi/2} \sin^6\theta\,\cos^2\theta\, d\theta$. Then the formula of Problem 108 gives the answer $\frac{5}{4}\pi$.

112. $\sqrt{1 + (dy/dx)^2} = \sqrt{1 + \sqrt{x}}$; the curve's length is $L = \displaystyle\int_0^1 \sqrt{1 + \sqrt{x}}\, dx$. Let $x = \tan^4 u$.

$$L = \int_0^{\pi/4} 4\sqrt{1 + \tan^2 u}\,\tan^3 u\,\sec^2 u\, du = 4\int_0^{\pi/4} \sec^3 u\,\tan^3 u\, du$$

$$= 4\int_0^{\pi/4}(\sec^3 u)(\sec^2 u - 1)\tan u\, du = 4\int_0^{\pi/4}(\sec^4 u - \sec^2 u)(\sec u\,\tan u)\, du$$

$$= \left[\tfrac{4}{5}\sec^5 u - \tfrac{4}{3}\sec^3 u\right]_0^{\pi/4} = \tfrac{4}{5}\left(4\sqrt{2} - 1\right) - \tfrac{4}{3}\left(2\sqrt{2} - 1\right) = \tfrac{8}{15}\left(1 + \sqrt{2}\right) \approx 1.28758.$$

115. Let $u = e^x$. $\displaystyle\int \frac{1}{1 + e^x + e^{-x}}\, dx = \int \frac{1}{u^2 + u + 1}\, du$.

118. With the recommended substitution, the numerator is $\frac{1}{2}\, du$, so

$$\int \frac{1 + 2x^2}{x^5(1 + x^2)^3}\, dx = \tfrac{1}{2}\int u^{-3}\, du = -\tfrac{1}{4}u^{-2} + C = -\frac{1}{4(x^4 + x^2)^2} + C.$$

Note: A partial fraction decomposition results in

$$\int \frac{1 + 2x^2}{x^5(1 + x^2)^3}\, dx = \frac{1}{2x^2} - \frac{1}{4x^4} - \frac{1}{2(x^2 + 1)} - \frac{1}{4(x^2 + 1)^2} + C.$$

121. Let $3x - 2 = u^2$: $\displaystyle\int x^3\sqrt{3x - 2}\, dx = \int\left(\frac{u^2 + 2}{3}\right)^3 \cdot u \cdot \tfrac{2}{3}u\, du$

124. Let $x - 1 = u^2$:

$$\int x^2(x - 1)^{3/2}\, dx = \int(2u^8 + 4u^6 + 2u^4)\, du = \tfrac{2}{9}(x - 1)^{9/2} + \tfrac{4}{7}(x - 1)^{7/2} + \tfrac{2}{5}(x - 1)^{5/2} + C$$

130. $\int \sqrt{1+\sqrt{x}}\; dx = \frac{4}{15}\left(3\sqrt{x}-2\right)\left(1+\sqrt{x}\right)^{3/2} + C.$

133. Area: $A = 2\int_0^1 x\sqrt{1-x}\; dx.$ For variety, let $x = \sin^2\theta.$

$$A = 2\int_0^{\pi/2} \left(\sin^2\theta\right)\left(\cos\theta\right)\left(2\sin\theta\,\cos\theta\right)\,d\theta$$

$$= 4\int_0^{\pi/2} \left(\cos^2\theta - \cos^4\theta\right)\sin\theta\,d\theta$$

$$= 4\left[\tfrac{1}{5}\cos^5\theta - \tfrac{1}{3}\cos^3\theta\right]_0^{\pi/2} = \tfrac{8}{15}.$$

136. $\int \dfrac{1}{5+4\cos\theta}\,d\theta = \int \dfrac{2}{u^2+9}\,du = \tfrac{2}{3}\arctan\left(\dfrac{u}{3}\right) + C = \tfrac{2}{3}\arctan\left(\tfrac{1}{3}\tan\dfrac{\theta}{2}\right) + C.$

139. $\int \dfrac{1}{\sin\theta+\cos\theta}\,d\theta = \int \dfrac{2}{1+2u-u^2}\,du = \dfrac{\sqrt{2}}{2}\ln\left|\dfrac{u+\sqrt{2}-1}{u-\sqrt{2}-1}\right| + C.$

142. $\int \dfrac{\sin\theta-\cos\theta}{\sin\theta+\cos\theta}\,d\theta = -2\int \dfrac{u^2+2u-1}{(u^2+1)(u^2-2u-1)}\,du$

$$= \ln\left|\dfrac{u^2+1}{u+\sqrt{2}-1}\right| - \ln\left|u-\sqrt{2}-1\right| + C = -\ln\left|\sin\theta+\cos\theta\right| + C.$$

Chapter 10: Polar Coordinates and Plane Curves
Section 10.1

1. The given line has slope $-\frac{1}{2}$.

4. By implicit differentiation, $\dfrac{dy}{dx} = \dfrac{1}{2y}$; in particular, the slope of the tangent at $(6, -3)$ is $-\frac{1}{6}$. So the tangent line has equation $y + 3 = -\frac{1}{6}(x - 6)$; that is, $x + 6y + 12 = 0$.

7. Write the equation in the form $x^2 + 2x + 1 + y^2 = 5$.

10. Write the equation in the form $x^2 + 8x + 16 + y^2 - 6y + 9 = 25$ to conclude that the center is at $(-4, 3)$ and the radius is 5.

13. Complete the squares as follows: $2x^2 + 2y^2 - 2x + 6y = 13$;
$$x^2 + y^2 - x + 3y = \tfrac{13}{2};$$
$$x^2 - x + \tfrac{1}{4} + y^2 + 3y + \tfrac{9}{4} = \tfrac{36}{4};$$
$$\left(x - \tfrac{1}{2}\right)^2 + \left(y + \tfrac{3}{2}\right)^2 = 3^2.$$
Thus the center is at $\left(\tfrac{1}{2}, -\tfrac{3}{2}\right)$ and the radius is 3.

16. Center: $\left(\tfrac{2}{3}, \tfrac{3}{2}\right)$; radius: 12.

19. Complete the squares as follows: $x^2 + y^2 - 6x - 10y + 84 = 0$;
$$x^2 - 6x + 9 + y^2 - 10y + 25 + 50 = 0;$$
$$(x - 3)^2 + (y - 5)^2 = -50.$$
It follows that there are no points on the graph.

22. The line has slope 1, so the radius to the point of tangency has slope -1. The radius is a segment of the line with equation $y + 2 = -(x - 2)$; that is, $x + y = 0$. The point of tangency is the simultaneous solution of the equations of the two lines: $x = -2$, $y = 2$—therefore they meet at $(-2, 2)$. The length of the radius is the distance from $(2, -2)$ to $(-2, 2)$, which is $4\sqrt{2}$. Therefore the equation of the circle is $(x - 2)^2 + (y + 2)^2 = 32$.

25. The squares of the distances are equal, so $P(x, y)$ satisfies $(x - 3)^2 + (y - x)^2 = (x - 7)^2 + (y - 4)^2$. Expand and simplify to obtain the answer.

28. The point $P(x, y)$ satisfies the equation $x + 3 = \sqrt{(x - 3)^2 + y^2}$. Expand and simplify to obtain $y^2 = 12x$: The locus is a parabola, opening to the right, with its vertex at the origin, and symmetric about the x-axis.

31. If $P(a, b)$ is the point of tangency, then $b = a^2$, and the slope of the tangent line can be measured in two different ways. They are equal, and we thereby obtain
$$\frac{1 - a^2}{2 - a} = 2a.$$
We find two solutions: $a = 2 - \sqrt{3}$ and $a = 2 + \sqrt{3}$. But $b = a^2$, so $b = 7 - 4\sqrt{3}$ or $b = 7 - 4\sqrt{3}$. Thus we obtain the two answers given in the text.

34. The second condition means that all such lines have slope 3. If such a line is tangent to the graph of $y = x^3$ at the point (a, a^3), its slope must also be $3a^2 = 3$. Thus $a = 1$ or $a = -1$. Thus there are two such lines:
$$y - 1 = 3(x - 1) \quad \text{(through $(1, 1)$ with slope 3)} \quad \text{and}$$
$$y + 1 = 3(x + 1) \quad \text{(through $(-1, -1)$ with slope 3)}.$$

Section 10.2

4. Answer: $r \sin \theta = 6$; a better form is $r = 6 \csc \theta, 0 < \theta < \pi$.

10. Answer: $r = \dfrac{4}{\sin \theta + \cos \theta}$. It would be best to restrict the values of θ to the range $-\pi/4 < \theta < 3\pi/4$.

13. Begin by multiplying each side of the equation by r.

16. First, $r^2 = 2r + r \sin \theta$, so $x^2 + y^2 = 2\sqrt{x^2 + y^2} + y$. Hence

$$\left(x^2 + y^2 - y\right)^2 = \left(2\sqrt{x^2 + y^2}\right)^2 = 4x^2 + 4y^2;$$
$$x^4 + 2x^2y^2 + y^4 - 2x^2y - 2y^3 + y^2 = 4x^2 + 4y^2;$$
$$x^4 + 2x^2y^2 + y^4 - 2x^2y - 2y^3 - 4x^2 - 3y^2 = 0.$$

22. $y = x - 2$; $r(\cos \theta - \sin \theta) = 2, \quad -5\pi/4 < \theta < \pi/4$

28. $(x - 5)^2 + (y + 2)^2 = 25$; $r^2 - 10r \cos \theta + 4r \sin \theta + 4 = 0$

31. Matches Fig. 10.2.18.

34. Matches Fig. 10.2.22.

37. Given $r = a \cos \theta + b \sin \theta$, multiply both sides by r and convert to rectangular coordinates to obtain, successively,

$$x^2 + y^2 = ax + by;$$
$$x^2 - ax + \tfrac{1}{4}a^2 + y^2 - by + \tfrac{1}{4}b^2 = \tfrac{1}{2}\left(a^2 + b^2\right);$$
$$\left(x - \frac{a}{2}\right)^2 + \left(y - \frac{b}{2}\right)^2 = \frac{a^2 + b^2}{4}.$$

Thus the graph is a circle with center $(a/2, b/2)$ and radius $\tfrac{1}{2}\sqrt{a^2 + b^2}$—unless $a = b = 0$, in which case the graph consists of the single point $(0, 0)$.

40. Multiply each side by r, then complete the square to obtain the equation

$$(x - 1)^2 + (y - 1)^2 = 2.$$

The graph is a circle with center at $(1, 1)$ and radius $\sqrt{2}$. It has no symmetries of the sort mentioned; its graph is shown on the right.

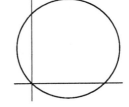

46. Given:

$$r^2 = 4 \cos 2\theta.$$

Note that there is no graph when $\cos 2\theta < 0$; that is, for $\pi/4 < \theta < 3\pi/4$ and for $5\pi/4 < \theta < 7\pi/4$. The graph is symmetric about both coordinate axes and about the origin, and is shown on the right.

52. Given:

$$r^2 = 4 \sin \theta.$$

The graph is symmetric about the x-axis, the y-axis, and the origin. It is shown at the right.

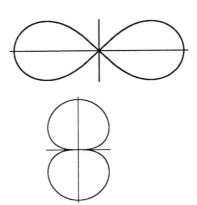

55. The graph of $r = \sin\theta$ is shown as a dashed curve in the figure to the right. The other graph—that of the equation $r = \cos 2\theta$—is shown as a solid curve. The Answer section of the text gives the polar coordinates of the four points of intersection.

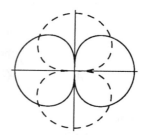

58. The graph of $r^2 = 4\sin\theta$ is shown as a dashed curve in the figure to the right and the graph of $r^2 = 4\cos\theta$ is a solid curve. The five intersection points are the origin and the four points for which $r = 8^{1/4}$ and θ is an odd multiple of $\pi/4$.

61. The given equation has polar form $a^2 r^2 = (r^2 - br\sin\theta)^2$. Expand the right-hand side and cancel r^2 to obtain

$$a^2 = r^2 - 2br\sin\theta + b^2\sin^2\theta = (r - b\sin\theta)^2.$$

It follows that $r = \pm a + b\sin\theta$, and hence the graph is a limaçon.

Section 10.3

1. The area is shown below:

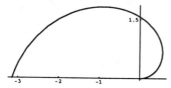

4. The area is shown below:

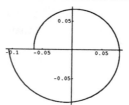

7. $A = 2\displaystyle\int_0^{\pi/2} \frac{1}{2}\left(4\cos^2\theta\right)\, d\theta.$

10. $A = \displaystyle\int_0^{2\pi} \frac{1}{2}(4)\left(1 - \sin\theta\right)^2\, d\theta = 2\int_0^{2\pi} \left(1 - 2\sin\theta + \frac{1}{2}\left(1 - \cos 2\theta\right)\right)\, d\theta$

$$= \left[2\theta + 4\cos\theta + \theta - \frac{1}{2}\sin 2\theta\right]_0^{2\pi} = 6\pi.$$

13. $A = \frac{1}{2}\displaystyle\int_0^{\pi} 16\cos^2\theta\, d\theta = 4\int_0^{\pi} \left(1 + \cos 2\theta\right)\, d\theta = 4\left[\theta + \frac{1}{2}\sin 2\theta\right]_0^{\pi} = 4\pi.$

16. $A = \frac{1}{2}\displaystyle\int_0^{2\pi} \left(10 + 6\sin\theta + 6\cos\theta + 2\sin\theta\,\cos\theta\right)\, d\theta = \frac{1}{2}\left[10\theta - 6\cos\theta + 6\sin\theta + \sin^2\theta\right]_0^{2\pi} = 10\pi.$

19. The graph of the given curve is shown on the right. The area within each of the eight loops is

$$A = 2\int_0^{\pi/8} \frac{1}{2}\left(4\cos^2 4\theta\right)\, d\theta.$$

22. The graph of $r^2 = 4\cos 2\theta$ is shown on the right. The area within each of the loops is

$$A = \tfrac{1}{2} \int_{-\pi/4}^{\pi/4} 4\cos 2\theta \, d\theta$$

$$= \Big[\sin 2\theta\Big]_{-\pi/4}^{\pi/4} = 2.$$

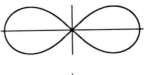

25. The two curves are shown on the right. The region in question has area

$$A = \tfrac{1}{2} \int_{\pi/6}^{5\pi/6} \left(4\sin^2\theta - 1\right) \, d\theta.$$

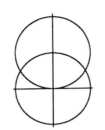

28. We are to find the area outside the circle $r = 2$ and within the curve $r = 2 + \cos\theta$; these curves are shown on the right. The curves meet where $\theta = \pi/2$ and where $\theta = -\pi/2$, so the area in question is

$$A = \tfrac{1}{2} \int_{-\pi/2}^{\pi/2} \left((2 + \cos\theta)^2 - 2^2\right) \, d\theta$$

$$= \int_0^{\pi/2} \left(4\cos\theta + \tfrac{1}{2}(1 + \cos 2\theta)\right) \, d\theta$$

$$= \Big[4\sin\theta + \tfrac{1}{2}\theta + \tfrac{1}{4}\sin 2\theta\Big]_0^{\pi/2} = \tfrac{1}{4}(\pi + 16).$$

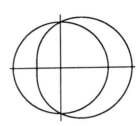

31. The curves, shown on the right, intersect where $\theta = \pi/8$. By symmetry, the area between them is

$$A = 4 \int_0^{\pi/8} \tfrac{1}{2}\sin 2\theta \, d\theta.$$

34. The two curves meet at the origin, at the point with Cartesian coordinates $(-2, 0)$, and at the two points where $\cos\theta = 3 - 2\sqrt{2}$. Let the least positive value of θ that satisfies the latter equation be denoted by ω. Then the area within the figure-8 curve but outside the cardioid is given by

$$A = \int_0^{\omega} \left(4\cos\theta - (1 - \cos\theta)^2\right) \, d\theta = \int_0^{\omega} \left(6\cos\theta - 1 - \tfrac{1}{2}(1 + \cos 2\theta)\right) \, d\theta$$

$$= \Big[6\sin\theta - \tfrac{3}{2}\theta - \tfrac{1}{4}\sin 2\theta\Big]_0^{\omega}$$

$$= 12\sqrt{3\sqrt{2} - 4} - \tfrac{3}{2}\omega - \left(3 - 2\sqrt{2}\right)\sqrt{3\sqrt{2} - 4} \approx 3.7289587.$$

37. The Cartesian equation of the curve is $\left(x - \tfrac{1}{2}\right)^2 + \left(y - \tfrac{1}{2}\right)^2 = \left(\tfrac{1}{2}\sqrt{2}\right)^2$. So the graph is a circle of radius $\tfrac{1}{2}\sqrt{2}$.

40. One way to solve this problem is to let one circle have polar equation $r = 2a\cos\theta$ and the other have polar equation $r = a$. The first circle then has center at $x = a$ on the x-axis and passes through the origin; the second has center at the origin and passes through the center of the first. They meet where $\theta = \pi/4$ and where $\theta = -\pi/4$, and the area within either that is outside the other is

$$A_1 = \int_0^{\pi/4} \left(4a^2\cos^2\theta - a^2\right) \, d\theta = a^2 \int_0^{\pi/4} \left(2(1 + \cos 2\theta) - 1\right) \, d\theta = a^2 \Big[\theta + \sin 2\theta\Big]_0^{\pi/4} = \tfrac{1}{4}a^2(\pi + 4).$$

Let A denote the area within both circles. Each circle has area πa^2, so

$$2\pi a^2 = (A_1 + A) + (A_1 + A). \quad \text{Therefore}$$
$$A = \pi a^2 - A_1 = \tfrac{1}{4}a^2(3\pi - 4).$$

43. The point of intersection in the second quadrant is located where $\theta = \alpha \approx 2.326839$. Using symmetry, the total area of the shaded region R is approximately

$$2\int_0^\alpha \tfrac{1}{2}(e^{-t/5})^2\, dt + 2\int_\alpha^\pi \tfrac{1}{2}\left[2(1+\cos t)\right]^2 dt \approx 1.58069.$$

Section 10.4

1. Solve the first equation for $t = x - 1$; substitute into the second: $y = 2(x-1) - 1 = 2x - 3$. Answer: $y = 2x - 3$.

4. First, $x^2 = t$, so $y = 3x^2 - 2$, $x \geq 0$.

7. From $x = e^t$ we may conclude that $y = 4e^{2t} = 4x^2$. So the answer is $y = 4x^2$, $x > 0$.

10. $1 + \sinh^2 t = \cosh^2 t$, hence because $\cosh t > 0$ for all t, we get $y = +\sqrt{1 + x^2}$.

16. $y = 1 - x^2$, $-1 \leq x \leq 1$.

22. We use the equation $\cot\psi = \dfrac{1}{r}\cdot\dfrac{dr}{d\theta}$ from the text. Here, $\dfrac{dr}{d\theta} = -\dfrac{1}{\theta^2}$. And because $r = 1/\theta$, $\cot\psi = (\theta)\left(\dfrac{-1}{\theta^2}\right) = -\dfrac{1}{\theta}$. When $\theta = 1$, $\cot\psi = -1$. Therefore $\psi = 3\pi/4$.

25. There are horizontal tangents at $(1,2)$ and $(1,-2)$. There is a vertical tangent line at $(0,0)$. There is no tangent line at the other x-intercept $(3,0)$ because the curve crosses itself with two different slopes there, namely the slopes $\pm\sqrt{3}$.

28. There are horizontal tangents at the four points $(\pm\tfrac{1}{2}\sqrt{6}, \pm\tfrac{1}{2}\sqrt{2})$. There are vertical tangents at $(\pm 2, 0)$ but no tangent at the third x-intercept $(0,0)$; at that point, the curve crosses itself with slopes $+1$ and -1.

31. If the slope at $P(x,y)$ is m, then $2y\dfrac{dy}{dx} = 4p$, so $y = \dfrac{2p}{dy/dx} = \dfrac{2p}{m}$. Therefore $4px = y^2 = \dfrac{4p^2}{m^2}$, and so $x = \dfrac{p}{m^2}$, $y = \dfrac{2p}{m}$.

34. $|OC| = a - b$. So C has coordinates $x = (a-b)\cos t$, $y = (a-b)\sin t$. The arc length from the point of tangency to $A(a,0)$ is the same as that to P; denote it by s. Note that $s = ta$. Let α be the angle OCP and θ the angle supplementary to α, so that $\theta = \pi - \alpha$. Then $s = b\theta$. So $ta = b\theta$. The radius b is at the angle $-(\theta - t) = t - \theta$ from the horizontal, so P has coordinates

$$x = (a-b)\cos t + b\cos(t-\theta), \quad y = (a-b)\sin t + b\sin(t-\theta).$$

Now $\theta = \dfrac{a}{b}t$, so $t - \theta = t - \dfrac{a}{b}t = \dfrac{b-a}{b}t$. Hence P has the coordinates given in the statement of the problem.

Section 10.5

1. Note first that $y = 2t^2 + 1$ is always positive, so the curve lies entirely above the x-axis. Moreover, as t goes from -1 to 1, $dx = 3t^2\, dt$ is positive, so the area is $A = \displaystyle\int_{-1}^1 (2t^2 + 1)(3t^2)\, dt = \tfrac{22}{5}$.

4. $A = \int_1^0 -3e^{2t}\,dt = \frac{3}{2}\left(e^2 - 1\right).$

7. $V = \int_{-1}^1 \pi\left(2t^2 + 1\right)^2\left(3t^2\right)\,dt.$

10. $V = \int_\pi^0 -\pi e^{2t} \sin t\,dt = \frac{\pi}{5}\left(e^{2\pi} + 1\right).$

13. $\left(\dfrac{dx}{dt}\right)^2 + \left(\dfrac{dy}{dt}\right)^2 = 2$, so $L = \int_{\pi/4}^{\pi/2} \sqrt{2}\,dt.$

16. $L = \int_{2\pi}^{4\pi} \sqrt{1 + \theta^2}\,d\theta$

$= 2\pi\sqrt{1 + 16\pi^2} + \frac{1}{2}\ln\left(4\pi + \sqrt{1 + 16\pi^2}\right) - \pi\sqrt{1 + 4\pi^2} - \frac{1}{2}\ln\left(2\pi + \sqrt{1 + 4\pi^2}\right),$

which is approximately equal to 59.563022.

19. $A = 2\int_0^1 2\pi t^3\sqrt{9t^4 + 4}\,dt.$

22. $A = \int_0^{\pi/2} 2\pi\left(e^\theta \cos\theta\right)\left(\sqrt{2}\,e^\theta\right)\,dt = \frac{2}{5}\pi\sqrt{2}\left(e^\pi - 2\right).$

25. **(a)** $A = 2\int_0^\pi (b\sin t)(a\sin t)\,dt.$ **(b)** $V = 2\int_0^\pi \left(b^2\sin^2 t\right)(a\sin t)\,dt.$

28. $A = \int_0^{2\pi} 2\pi a\left(b + a\cos t\right)\,dt = 4\pi^2 ab.$

31. Here, $ds = 3a\,|\sin t \cos t|\,dt$, so $A = (4\pi)(3a)\int_0^{\pi/2} (\sin t \cos t)\left(a\sin^3 t\right)\,dt.$

34. $ds = \left(t^2 + 3\right)dt$, so the arc length is $L = \int_{-3}^3 \left(t^2 + 3\right)dt = 36.$

40. **(b)** First, the arc-length element is $ds = \sqrt{[x'(t)]^2 + [y'(t)]^2}\,dt = \sqrt{a^2 t^2}\,dt = |at|\,dt = at\,dt.$ So the length of the involute from $t = 0$ to $t = \pi$ is

$$\int_{t=0}^\pi 1\,ds = \int_0^\pi at\,dt = \frac{1}{2}\pi^2 a.$$

43. Given $r(t) = 3\sin 3\theta$, remember that roses with *odd* coefficients are swept out *twice* in the interval $0 \leq \theta \leq 2\pi$, so we should integrate $ds = \sqrt{[r(\theta)]^2 + [r'(\theta)]^2}\,d\theta = \sqrt{45 + 36\cos 6\theta}\,d\theta$ from 0 to π to obtain the total length of the rose:

$$\int_{\theta=0}^\pi 1\,ds = \int_0^\pi \sqrt{45 + 36\cos 6\theta}\,d\theta \approx 20.0473.$$

46. When the rose of Problem 45 is rotated around the x-axis, the entire surface is generated twice. To obtain each part of the surface once, we will rotate the part of the rose from $\theta = 0$ to $\theta = \pi/4$ and, separately, the part from $\theta = \pi/4$ to $\pi/2$. We will set up an integral for each surface area, add the results, and double the sum. With $x(\theta) = r(\theta)\sin\theta$ and the arc length element ds of Problem 45, we get the integrals

$$\int_{\theta=0}^{\pi/4} 2\pi y\,ds \approx 5.46827 \quad \text{and} \quad \int_{\theta=\pi/4}^{\pi/2} (-2\pi y)\,ds \approx 16.1232,$$

for a total area of approximately 43.1829.

49. Given: $r(\theta) = \cos(7\theta/3)$. To sweep out all seven "petals" of this quasi-rose, you need to let θ vary from 0 to 3π. The length of the graph is

$$\int_0^{3\pi} \sqrt{\tfrac{1}{9}(29 - 20\cos(14\theta/3))}\ d\theta \approx 16.3428.$$

52. Now the curve of Problems 50 and 51 is to be rotated around the y-axis. We will use the same part of the curve (the part in the first quadrant) and double the answer.
(a) The surface area generated is

$$2\int_0^{\pi/2} 2\pi x(t)\sqrt{\cos^2 t + 4\cos^2 2t}\ dt \approx 17.7205.$$

(b) Using the method of cylindrical shells, the volume enclosed by that surface is

$$2\int_0^{\pi/2} 2\pi x(t)y(t)x'(t)\,dt = 4\pi \int_0^{\pi/2} 2\sin^2 t \cos^2 t\,dt$$

$$= 4\pi \int_0^{\pi/2} \tfrac{1}{2}(1 - \cos^2 2t)\,dt = 2\pi \int_0^{\pi/2}\left(1 - \frac{1 + \cos 4t}{2}\right) dt$$

$$= 2\pi \int_0^{\pi/2} \left(\tfrac{1}{2} - \tfrac{1}{2}\cos 4t\right)\,dt = 2\pi \left[\tfrac{1}{2}t - \tfrac{1}{8}\sin 4t\right]_0^{\pi/2} = \tfrac{1}{2}\pi^2.$$

Section 10.6

1. The value of p is 3, so the parabola has equation $y^2 = 12x$.

4. The value of p is 2, so this parabola has equation $(y + 1)^2 = -8(x + 1)$. Its vertex is at the point $(-1, -1)$ and it opens to the left.

7. Here, we have $p = \tfrac{3}{2}$, so this parabola has equation $x^2 = -6\left(y + \tfrac{3}{2}\right)$.

10. Here we have $p = 1$, so this parabola has equation

$$(y - 1)^2 = 4(x + 3)$$

and its graph is shown at the right.

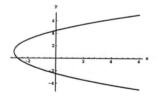

16. The given equation may be written in the form

$$y^2 + 6y + 9 = 2x - 6;$$
$$(y + 3)^2 = (4)(\tfrac{1}{2})(x - 3).$$

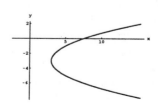

So $p = \tfrac{1}{2}$; vertex $(3, -3)$; directrix $x = \tfrac{5}{2}$; focus at $\left(\tfrac{7}{2}, -3\right)$; axis the horizontal line $y = -3$. The graph is on the right.

19. Note that this is a maximum-minimum problem, and recall that a distance is minimized when its square is minimized.

22. Set up coordinates so that the parabola has vertex $V(-p, 0)$. Then the equation of the comet's orbit is

$$y^2 = 4p(x + p).$$

The line $y = x$ meets the orbit of the comet at the point (a, b) (say), which is $100\sqrt{2}$ million miles from the origin (which is also where both the sun and the focus of the parabola are located). Therefore

$$a^2 = 4p(a + p) \quad \text{and} \quad \sqrt{a^2 + a^2} = \left(100\sqrt{2}\right)(10^6) = 10^8\sqrt{2}.$$

It follows that $a = 10^8$. Next, $a^2 = 4p(a + p)$. We apply the quadratic formula to find that $p = \frac{1}{2}(\sqrt{2} - 1)(10^8)$. The vertex is at distance p from the focus; therefore, by the result of Problem 19, the closest approach is approximately 20.71 million miles.

28. **(a)** $\alpha = \pi/6$: Range: $\dfrac{2500}{9.8} \cdot \dfrac{\sqrt{3}}{2} \approx 220.9$ meters

Time aloft: $T = \dfrac{2v_0 \sin \alpha}{3} = \dfrac{100(0.5)}{9.8} \approx 5.102$ seconds.

(b) $\alpha = \pi/3$: Range: $\dfrac{2500}{9.8} \cdot \dfrac{\sqrt{3}}{2} \approx 220.9$ meters

Time aloft: $T = \dfrac{100}{9.8} \cdot \dfrac{\sqrt{3}}{2} \approx 8.837$ seconds.

Section 10.7

1. The location of the vertices makes it clear that the center of the ellipse is at $(0, 0)$. Therefore its equation may be written in the standard form

$$\left(\frac{x}{4}\right)^2 + \left(\frac{y}{5}\right)^2 = 1.$$

4. We immediately have $a = 6$ and $b = 4$, so the equation is

$$\left(\frac{x}{4}\right)^2 + \left(\frac{y}{6}\right)^2 = 1.$$

7. First, $a = 10$ and $e = \frac{1}{2}$. So $c = ae = 5$ and thus $b^2 = 100 - 25 = 75$. So the equation is

$$\frac{x^2}{100} + \frac{y^2}{75} = 1.$$

10. First, $c = 4$ and $9 = c/e^2$; therefore $e = \frac{2}{3}$. So $a = c/e = 6$; $b^2 = a^2 - c^2 = 20$. Equation:

$$\frac{x^2}{20} + \frac{y^2}{36} = 1.$$

13. "Move" the center first to the origin to obtain the equation $\dfrac{x^2}{25} + \dfrac{y^2}{16} = 1$. Then apply the translation principle to obtain the answer.

16. In standard form, the equation of this ellipse is

$$\frac{x^2}{4} + \frac{y^2}{16} = 1.$$

So the major axis is vertical, $b = 2$, $a = 4$, the minor axis is of length 4, and the major axis is of length 8. The center is at the origin. Foci: $(0, -2\sqrt{3})$ and $(0, 2\sqrt{3})$. Its graph is shown at the right.

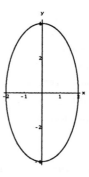

19. The equation can be put in the standard form $\left(\dfrac{x}{2}\right)^2 + \left(\dfrac{y-4}{3}\right)^2 = 1$. The rest is routine.

22. In the usual notation, we have $2a = 0.467 + 0.307 = 0.774$. So $a = 0.387$, $e = 0.206$. Therefore $c = ae = 0.079722$, and
$$b = \sqrt{a^2 - c^2} \approx 0.378699621;$$

we'll use $b = 0.3787$. Therefore the ellipse has major axis 0.774, minor axis 0.7574; in terms of percentages, a is about 2.2% greater than b. Is this a nearly circular orbit? Decide for yourself: Compare the circle (on the left) below, with diameter 0.7657, with the ellipse (on the right) below with the shape of the orbit of the planet.

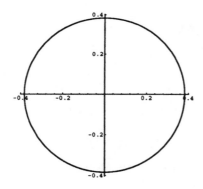
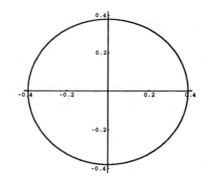

28. Begin with the equation
$$\sqrt{(x-3)^2 + (y+3)^2} + \sqrt{(x+3)^2 + (y-3)^2} = 10.$$
This leads to the equation $16x^2 + 18xy + 16y^2 = 175$.

Section 10.8

1. It follows that $c = 4$, $a = 1$, and $b^2 = 15$. So the standard equation is
$$x^2 - \frac{1}{15}y^2 = 1.$$

4. First, $a = 3$, and $b/a = 3/4 = b/3$, so $b = 9/4$. So the equation of the hyperbola is
$$\tfrac{1}{9}x^2 - \tfrac{16}{81}y^2 = 1.$$

7. Let us interchange x and y in the data given in the problem, then restore their meanings with a second interchange at the last step. That is, we assume the foci to be at $(-6, 0)$ and $(6, 0)$, and that the eccentricity is $e = 2$. It follows that $c = 6$, and thus that $a = c/e = 3$. Thus $b^2 = c^2 - a^2 = 27$, and this hyperbola would have equation $x^2/9 - y^2/27 = 1$. Now we interchange x and y to obtain the correct answer.

10. We have $c = 9$; also, $4 = c/e^2$, so $e = 3/2$. Next, $a = c/e = 6$, so $b^2 = c^2 - a^2 = 45$. Hence the equation is

$$\frac{y^2}{36} - \frac{x^2}{45} = 1.$$

13. We "move" the data so that the center is at $(0,0)$: Replace x with $x - 1$ and y with $y + 2$. The translated hyperbola has vertices at $(0, 3)$ and $(0, -3)$ and asymptotes $3x = 2y$ and $3x = -2y$. Now let's interchange x and y. One asymptote then has equation $y = 2x/3$, so $b/a = 2/3$. But $a = 3$, so $b = 2$. The equation of this hyperbola is $x^2/9 - y^2/4 = 1$. But we must interchange x and y again and then replace y with $y + 2$ and x with $x - 1$; this gives the answer.

16. Write $(x + 2)^2 - 2y^2 = 4$, thus $\dfrac{(x + 2)^2}{4} - \dfrac{y^2}{2} = 1$. So $a = 2$, $b = \sqrt{2}$, and $c^2 = a^2 + b^2 = 6$. Therefore the center is at the point $(-2, 0)$, the foci are at $(-2 + \sqrt{6}, 0)$ and $(-2 - \sqrt{6}, 0)$, and the asymptotes have the equations

$$y = \left| \tfrac{1}{2}\sqrt{2}\,(x + 2) \right|.$$

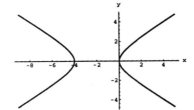

The graph is shown at the right.

22. First show that the equation of the tangent at $P_0 (x_0, y_0)$ is

$$y - y_0 = \left(\frac{b^2 x_0}{a^2 y_0} \right)(x - x_0).$$

If $y_0 = 0$, use dx/dy rather than dy/dx to get the desired result.

25. Note that $a^2 = \frac{9}{2} + \frac{9}{2}$, so that $a = 3$; therefore $2a = 6$. Then

$$\sqrt{(x - 5)^2 + (y - 5)^2} + 6 = \sqrt{(x + 5)^2 + (y + 5)^2}.$$

Now apply either the technique or the formula of the solution of Problem 24.

Chapter 10 Miscellaneous

4. Given: $y^2 = 4(x + y)$. Thus $y^2 - 4y + 4 = 4x + 4 = 4(x + 1)$; it follows that

$$(y - 2)^2 = 4(x + 1).$$

So the graph is a parabola; it has directrix $x = -2$, axis $y = 2$, vertex at $(-1, 2)$, focus at $(0, 2)$, and it opens to the right.

10. The given equation can be written in the form

$$\frac{(y - 1)^2}{4} - \frac{(x + 1)^2}{9} = 1.$$

So the graph is a hyperbola with center $(-1, 1)$. Because $c = \sqrt{13}$, the foci are at the points

$$\left(-1, 1 + \sqrt{13}\right) \quad \text{and} \quad \left(-1, 1 - \sqrt{13}\right).$$

The vertices are at $(-1, 3)$ and $(-1, -1)$. The transverse axis is vertical, of length 4, and the conjugate axis is horizontal, of length 6. The eccentricity is $\frac{1}{2}\sqrt{13}$, the directrices are

$$y = 1 + \tfrac{4}{13}\sqrt{13} \text{ and } y = 1 - \tfrac{4}{13}\sqrt{13},$$

and the asymptotes have the equations $3y = 2x + 5$ and $3y = -2x + 1$. The graph is shown on the right.

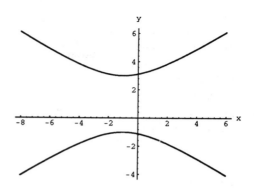

13.　The equation can be written in the form

$$\frac{(x-1)^2}{4} - \frac{y^2}{9} = 1,$$

so this is the equation of a hyperbola with center at $(1, 0)$ and horizontal transverse axis.

16.　This equation can be written in the form

$$\left(\frac{x-1}{2}\right)^2 + (y-2)^2 = 1,$$

so this is an equation of the ellipse with center $(1, 2)$, horizontal major axis of length 4, and vertical minor axis of length 2.

22.　Multiply each side by r to obtain $y^2 = x$. This is the equation of a parabola with axis the x-axis, opening to the right, with vertex $(0, 0)$, directrix $x = -\frac{1}{4}$, and focus $\left(\frac{1}{4}, 0\right)$.

28.　Given: $r = \dfrac{1}{1 + \cos\theta}$. In Cartesian coordinates we obtain $y^2 = -2x + 1$, so it is a parabola with focus at $(0, 0)$, directrix $x = 1$, and the vertex is at $\left(0, \frac{1}{2}\right)$. It opens to the left and its axis is the x-axis.

31.　The region whose area is sought is shown at the right.

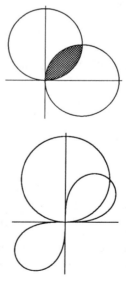

34.　The two regions are shown at the right. The curves cross at the points where $\sin\theta = 0$ and where $\cos\theta = \sin\theta$. We obtain the two solutions $r = 0$ and $r = \sqrt{2}$, $\theta = \pi/4$. The area of the small region is

$$A_S = \tfrac{1}{2} \int_0^{\pi/4} \left((2\sin 2\theta) - (2\sin\theta)^2 \right) d\theta,$$

which turns out to be $\frac{1}{4}(4 - \pi) \approx 0.214602$. The area of the large region is

$$A_L = \int_0^{\pi/2} \sin 2\theta \, d\theta = 1.$$

Therefore the total area outside the circle but within the lemniscate is $\frac{1}{4}(8 - \pi) \approx 1.214602$.

37. The circle and the cardioid are shown at the right. They intersect only at the pole.

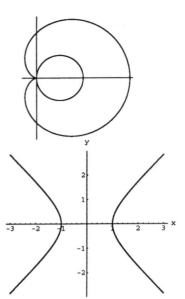

40. $x^2 - y^2 = 1$: Hyperbola;

$$\text{center at } (0,0),$$
$$\text{vertices at } (-1,0) \text{ and } (1,0).$$

The graph is shown at the right.

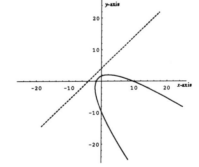

46. $dy/dx = -e^{-t}/e^t = -e^{-2t}$. When $t = 0$, we have $x = 1$, $y = 1$, and $dy/dx = -1$. One equation of the tangent line is therefore $x + y = 2$.

52. $A = \displaystyle\int_0^1 \sinh^2 t\, dt = \dfrac{e^4 - 4e^2 - 1}{8e^2} \approx 0.4067151$.

58. $dx/dt = 2t$, $dy/dt = 3$. So $ds = \sqrt{9 + 4t^2}\, dt$.

$$A = \int_{t=0}^2 2\pi y\, ds = \int_0^2 (2\pi)(3t)\sqrt{9 + 4t^2}\, dt = \left[\left(\frac{3\pi}{4}\right)\left(\frac{2}{3}\right)(9 + 4t^2)^{3/2}\right]_0^2 = 49\pi.$$

70. From the figure at the right, we see that matters would be greatly simplified if we were to rotate 45° counterclockwise to obtain the situation shown in the following figure. Then the parabola in the second figure has equation

$$r = \frac{2^{3/2}}{1 - \cos\theta}.$$

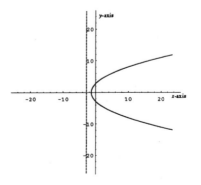

Therefore the parabola in the first figure has equation

$$r = \frac{2^{3/2}}{1 - \cos(\theta + (\pi/4))}.$$

After a considerable amount of algebra we find that we can write the Cartesian equation of the first parabola in the form

$$x^2 + 2xy + y^2 - 8x + 8y - 16 = 0.$$

76. We use the figure below. Note that we introduce a uv-coordinate system; in the rest of this discussion, all coordinates will be uv-coordinates.

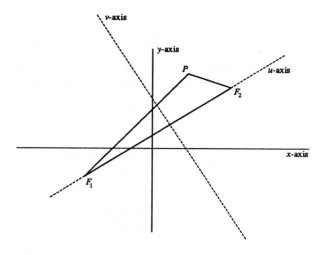

Choose the new axes so that $F_1 = F_1(c, 0)$ and $F_2 = F_2(-c, 0)$, $c > 0$. Suppose that $P = P(u, v)$. Then

$$|PF_2| = 2a + |PF_1|,$$

and therefore

$$\sqrt{(u + c)^2 + v^2} = 2a + \sqrt{(u - c)^2 + v^2}.$$

Consequently

$$(u + c)^2 + v^2 = 4a^2 + 4a\sqrt{(u - c)^2 + v^2} + (u - c)^2 + v^2.$$

Successive simplifications produce

$$4uc - 4a^2 = 4a\sqrt{(u - c)^2 + v^2};$$
$$uc - a^2 = a\sqrt{(u - c)^2 + v^2};$$
$$u^2c^2 - 2a^2uc + a^4 = a^2u^2 - 2a^2uc + a^2c^2 + a^2v^2;$$
$$u^2c^2 - a^2u^2 - a^2v^2 = a^2c^2 - a^4;$$
$$u^2\left(c^2 - a^2\right) - a^2v^2 = a^2\left(c^2 - a^2\right).$$

Now $|F_1F_2| > 2a$, so $c > a$. Thus $c^2 - a^2 = b^2$ for some $b > 0$. Hence

$$b^2u^2 - a^2v^2 = a^2b^2; \quad \text{that is,}$$
$$\frac{u^2}{a^2} - \frac{v^2}{b^2} = 1.$$

Therefore the locus of $P(u, v)$ is a hyperbola with vertices $(a, 0)$ and $(-a, 0)$ and foci $(c, 0)$ and $(-c, 0)$ (because $c^2 = a^2 + b^2$), and therefore the hyperbola has foci F_1 and F_2. Finally, if a circle with radius r_2 is centered at F_2 and another with radius r_1 is centered at F_1, with r_2 and r_1 satisfying the equation $r_2 = 2a + r_1$, then the two circles will intersect at a point on the hyperbola. You may thereby construct by straightedge-and-compass methods as many points lying on the hyperbola as you please.

79. The equation may be written in the form

$$(x+2)^2 - \tfrac{1}{3}y^2 = 1,$$

so the graph is a hyperbola with a vertical directrix. In the notation at the beginning of Section 10.8, we have $a = 1$ and $b = \sqrt{3}$, so $c^2 = a^2 + b^2 = 4$, and therefore $c = 2$. The eccentricity is then $e = c/a = 2$.

82. Here, the loop has polar equation

$$r = \frac{5\cos^2\theta \, \sin^2\theta}{\cos^5\theta + \sin^5\theta} \quad \text{for} \ \ 0 \le \theta \le \pi/2.$$

Therefore the area it bounds is

$$A = 25 \int_0^{\pi/4} \frac{\cos^4\theta \, \sin^4\theta}{\left(\cos^5\theta + \sin^5\theta\right)^2} \, d\theta.$$

The substitution $u = \tan\theta$ transforms this integral into

$$A = 25 \int_0^1 \frac{u^4}{\left(1 + u^5\right)^2} \, du = \tfrac{5}{2}.$$

Chapter 11: Infinite Series

Section 11.2

4. $a_n = \dfrac{(-1)^{n+1}}{2^n}$

10. $\lim\limits_{n\to\infty} \dfrac{(1/n^2) - 1}{(2/n^2) + 3} = -\frac{1}{3}$.

13. Show that if $|x| < 1$, then $x^n \to 0$ as $n \to +\infty$.

16. $0 \le 1 + (-1)^n \le 2$ for all n, so $0 \le \dfrac{1 + (-1)^n}{\sqrt{n}} \le \dfrac{2}{\sqrt{n}}$ for all n. So by the squeeze law, the limit is 0.

22. $a_n = n \cos \pi n = n$ if n is even, $-n$ if n is odd. Because $\{a_n\}$ has no upper bound (and no lower bound), it therefore diverges.

28. Let $x = 1/n$. Then $\lim\limits_{n\to\infty} a_n = \lim\limits_{x\to 0^+} \dfrac{\sin x}{x} = 1$.

34. $\ln a_n = \dfrac{1}{n} \ln(2n + 5)$ and $\lim\limits_{x\to\infty} \dfrac{1}{x} \ln(2x + 5) = 0$. Therefore $\lim\limits_{n\to\infty} a_n = e^0 = 1$.

37. Use the method of solution of Problem 26.

40. $\lim\limits_{x\to\infty} \ln\left((x^2 + 1)^{1/x}\right) = \lim\limits_{x\to\infty} \dfrac{\ln(x^2 + 1)}{x} = \lim\limits_{x\to\infty} \dfrac{2x}{x^2 + 1} = 0$. Therefore $\lim\limits_{n\to\infty} (n^2 + 1)^{1/n} = 1$, and therefore $\lim\limits_{n\to\infty} (-1)^n (n^2 + 1)^{1/n}$ does not exist.

43. $\lim\limits_{n\to\infty} \dfrac{n - 2}{n + 13} = \lim\limits_{n\to\infty} \dfrac{1 - \dfrac{2}{n}}{1 + \dfrac{13}{n}} = \dfrac{\left(\lim\limits_{n\to\infty} 1\right) - 2\cdot\left(\lim\limits_{n\to\infty}\dfrac{1}{n}\right)}{\left(\lim\limits_{n\to\infty} 1\right) + 13\cdot\left(\lim\limits_{n\to\infty}\dfrac{1}{n}\right)} = \dfrac{1 - 2\cdot 0}{1 + 13\cdot 0} = 1$.

46. The limit is

$$\lim_{n\to\infty} \left(\frac{n^3 - 5}{8n^3 + 7n}\right)^{1/3} = \left(\lim_{n\to\infty} \frac{n^3 - 5}{8n^3 + 7n}\right)^{1/3}$$

$$= \left(\lim_{n\to\infty} \frac{1 - \dfrac{5}{n^3}}{8 + \dfrac{7}{n^2}}\right)^{1/3} = \left(\frac{\left(\lim\limits_{n\to\infty} 1\right) - 5\cdot\left(\lim\limits_{n\to\infty}\dfrac{1}{n^3}\right)}{\left(\lim\limits_{n\to\infty} 8\right) + 7\cdot\left(\lim\limits_{n\to\infty}\dfrac{1}{n^2}\right)}\right)^{1/3}$$

$$= \left(\frac{1 - 5\cdot 0}{8 + 7\cdot 0}\right)^{1/3} = \left(\tfrac{1}{8}\right)^{1/3} = \tfrac{1}{2}.$$

49. Because $\lim\limits_{n\to\infty} \dfrac{n - 1}{n + 1} = 1$, and because $f(x) = 4\arctan x$ is continuous at $x = 1$, the limit is $4\arctan 1 = \pi$.

52. To say that

$$\lim_{n\to\infty} a_n = +\infty$$

means that, for every interval of the form $(c, +\infty)$, there exists a positive integer N such that, if $n \ge N$, then a_n is in the interval $(c, +\infty)$. If $\{a_n\}$ is an unbounded increasing sequence, then, no matter how large the number c, $a_k > c$ for some integer k. But then $a_n > c$ for all $n \ge k$, so that a_n is in the interval $(c, +\infty)$ for all $n \ge k$. This is what $\lim\limits_{n\to\infty} a_n = +\infty$ means.

55. **(b)** $G_1 = G_2 = G_3 = 1$ and $G_{n+1} = G_n + G_{n-2}$ for $n \ge 3$. The first few terms of the sequence are

$$1,\ 1,\ 1,\ 2,\ 3,\ 4,\ 6,\ 9,\ 13,\ 19,\ 28,\ 41,\ 60,\ 88,\ 129,\ 189,\ 277,\ 406,\ 595,\ \text{and } 872.$$

58. Much as in Example 13, you can show that $\{a_n\}$ is a bounded (by 2) increasing sequence, and therefore converges to [say] L. Then, again as in Example 13,

$$L = \lim_{n \to \infty} a_{n+1} = \lim_{n \to \infty} \sqrt{2 + a_n} = \sqrt{2 + L}.$$

Hence $L^2 = L + 2$, and it follows that $L = -1$ or $L = 2$. It is clear that $L > 0$, and therefore $a_n \to 2$ as $n \to \infty$.

Section 11.3

1. Geometric, ratio $\frac{1}{3}$, first term 1; sum $\dfrac{1}{1 - \frac{1}{3}} = \frac{3}{2}$.

4. $\lim_{n \to \infty} (0.5)^{1/n} = 1 \neq 0$: This series diverges by the nth -term test.

7. Geometric, ratio $\frac{1}{3}$.

10. $\lim_{x \to \infty} x^{1/x} = 1 \neq 0$. This series diverges by the nth -term test.

13. Geometric, ratio $r = -\dfrac{3}{e}$. The series diverges because $|r| > 1$.

16. $\displaystyle\sum_{n=0}^{\infty} \frac{1}{2^n} = 1$. So $\displaystyle\sum_{n=1}^{k} \left(\frac{2}{n} - \frac{1}{2^n} \right) > \left(\sum_{n=1}^{k} \frac{2}{n} \right) - 1$ for all positive integers k.

Because the harmonic series diverges to $+\infty$, so does $\displaystyle\sum_{n=1}^{\infty} \frac{2}{n}$, and therefore so does the given series.

19. $\dfrac{\frac{1}{5}}{1 - \frac{1}{5}} - \dfrac{\frac{1}{7}}{1 - \frac{1}{7}} = \frac{1}{4} - \frac{1}{6} = \frac{1}{12}$.

22. The series is geometric with ratio $r = \dfrac{\pi}{e}$, and it diverges because $|r| > 1$.

28. Diverges by the nth -term test ($2^{1/n} \to 1$ as $n \to +\infty$).

31. The series diverges by the nth-term test for divergence, because $\lim_{n \to \infty} \dfrac{n^2 - 1}{3n^2 + 1} = \frac{1}{3} \neq 0$.

34. This is a geometric series with nonzero first term and ratio $\arcsin 1$; the latter is approximately $1.5707963 > 1$, so this series diverges.

40. $0.252525\ldots = \frac{1}{4} \left(1 + 10^{-2} + 10^{-4} + 10^{-6} + \ldots \right) = \frac{1}{4} \left(\dfrac{1}{1 - \frac{1}{100}} \right) = \frac{25}{99}$.

46. This series is geometric with ratio $x - 1$, so it converges if $|x - 1| < 1$; that is, if $0 < x < 2$. For such x, its sum is $(x - 1)/(2 - x)$.

49. This series is geometric with ratio $5x^2/(x^2 + 16)$, so it converges if $5x^2/(x^2 + 16) < 1$; that is, if $-2 < x < 2$. For such x, its sum is $5x^2/(16 - 4x^2)$.

52. The nth partial sum is $\ln(n + 1)$, which approaches $+\infty$ as $n \to +\infty$. So this series diverges to $+\infty$.

58. The nth partial sum is $\dfrac{1}{2} - \dfrac{1}{n + 1} + \dfrac{1}{n + 2}$, which approaches $\frac{1}{2}$ as $n \to +\infty$.

64. Total distance traveled:

$$\begin{aligned}
D &= a + 2ra + 2r^2 a + 2r^3 a + 2r^4 a + \ldots \\
&= -a + 2a \left(1 + r + r^2 + r^3 + r^4 + \ldots \right) \\
&= -a + \frac{2a}{1 - r} = \frac{-a + ar + 2a}{1 - r} = a\frac{1 + r}{1 - r}.
\end{aligned}$$

67. $M_1 = (0.95) M_0$, $M_2 = (0.95) M_1 = (0.95)^2 M_0$, and so on. Therefore $M_n = (0.95)^n M_0$.

70. Let X denote 1, 2, 3, 4, or 5. Peter (who goes first) has these winning patterns and their probabilities:

6	$\dfrac{1}{6}$
X X X 6	$\dfrac{5^3}{6^4}$
X X X X X X 6	$\dfrac{5^6}{6^7}$
X X X X X X X X X 6	$\dfrac{5^9}{6^{10}}$
$\vdots$	$\vdots$

So his probability of winning the game is the sum of the geometric series with first term $\frac{1}{6}$ and ratio $\frac{125}{216}$, and so his probability of winning the game is $\frac{36}{91} \approx 0.3956$. Paul (who is second) wins with probability $\frac{30}{91} \approx 0.3297$ and Mary wins with probability $\frac{25}{91} \approx 0.2747$.

Section 11.4

4. $\quad \dfrac{1}{1-x} = 1 + x + x^2 + x^3 + x^4 + \dfrac{x^5}{(1-z)^6}$ for some z between 0 and x.

10. $f(x) = -7 + 5x - 3x^2 + x^3$.

16. $\tan x = 1 + \dfrac{2}{1!}\left(x - \dfrac{\pi}{4}\right) + \dfrac{4}{2!}\left(x - \dfrac{\pi}{4}\right)^2 + \dfrac{16}{3!}\left(x - \dfrac{\pi}{4}\right)^3 + \dfrac{80}{4!}\left(x - \dfrac{\pi}{4}\right)^4 + \dfrac{Q(z)}{5!}\left(x - \dfrac{\pi}{4}\right)^5$

for some z between $\pi/4$ and x, where $Q(z) = 16\sec^6 z + 88\sec^4 z \tan^2 z + 16\sec^2 z \tan^4 z$.

22. $1 + 2x + \dfrac{1}{2!}4x^2 + \dfrac{1}{3!}8x^3 + \dfrac{1}{3!}16x^4 + \cdots$

25. $\sin 2x = 2x - \dfrac{(2x)^3}{3!} + \dfrac{(2x)^5}{5!} - \dfrac{(2x)^7}{7!} + \dfrac{(2x)^9}{9!} - \cdots$

$\qquad = \displaystyle\sum_{n=0}^{\infty} \dfrac{(-1)^n 2^{2n+1} x^{2n+1}}{(2n+1)!} = 2x - \dfrac{4x^3}{3} + \dfrac{4x^5}{15} - \dfrac{8x^7}{315} + \dfrac{4x^9}{2835} - \cdots.$

28. $\sin^2 x = \frac{1}{2}(1 - \cos 2x)$

$\qquad = \frac{1}{2}\left(1 - \left[1 - \dfrac{x^2}{2!} + \dfrac{x^4}{4!} - \dfrac{x^6}{6!} + \cdots\right]\right)$

$\qquad = \dfrac{x^2}{2!2} - \dfrac{x^4}{4!2} + \dfrac{x^6}{6!2} - \dfrac{x^8}{8!2} + \cdots = \displaystyle\sum_{n=0}^{\infty} \dfrac{(-1)^n x^{2n+2}}{(2n+2)!2}.$

34. $1 + \dfrac{2x}{1!} + \dfrac{4x^2}{2!} + \dfrac{8x^3}{3!} + \dfrac{16x^4}{4!} + \cdots$

40.

n	$f^{(n)}(x)$	$f^{(n)}(0)$
0	$(1+x)^{1/2}$	1
1	$\dfrac{1}{2}(1+x)^{-1/2}$	$\dfrac{1}{2}$
2	$-\dfrac{1}{4}(1+x)^{-3/2}$	$-\dfrac{1}{4}$
3	$\dfrac{3}{8}(1+x)^{-5/2}$	$\dfrac{3}{8}$
4	$-\dfrac{(5)(3)}{2^4}(1+x)^{-7/2}$	$\dfrac{15}{16}$
$\vdots$	$\vdots$	$\vdots$

For large n, then, $f^{(n)}(0) = \dfrac{(2n-3)(2n-5)\cdots(5)(3)(1)}{2^n}(-1)^{n+1}$ and the answer is

$$1 + \frac{1}{1!\,2^1}\,x - \frac{1}{2!\,2^2}\,x^2 + \frac{3}{3!\,2^3}\,x^3 - \frac{5\cdot 3}{4!\,2^4}\,x^4 + \cdots .$$

46. Here is a plot of $\sin x$ and its Taylor polynomials, center zero, of degrees 3, 5, and 7.

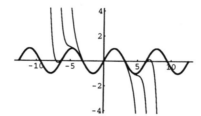

49. Here is a plot of $1/(1+x)$ and its Taylor polynomials, center zero, of degrees 2, 3, and 4.

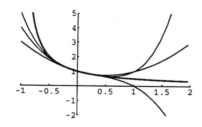

52. Given: $\alpha = \tan^{-1}(1/5)$.

(a) $\tan 2\alpha = \dfrac{\frac{1}{5}+\frac{1}{5}}{1-\frac{1}{25}} = \dfrac{5}{12}.$

(b) $\tan 4\alpha = \dfrac{\frac{5}{12}+\frac{5}{12}}{1-\frac{25}{144}} = \dfrac{120}{119}.$

(c) $\tan\left(\dfrac{\pi}{4}-4\alpha\right) = \dfrac{1-\frac{120}{119}}{1+\frac{120}{119}} = -\dfrac{1}{239}.$

(d) $\tan\left(\dfrac{\pi}{4}-4\alpha\right) = -\dfrac{1}{239};$

$\dfrac{\pi}{4}-4\alpha = -\arctan\dfrac{1}{239};$

$4\arctan\dfrac{1}{5} - \arctan\dfrac{1}{239} = \dfrac{\pi}{4}.$

Section 11.5

1. $\int_0^\infty \frac{x}{x^2+1}\, dx = \left[\frac{1}{2}\ln\left(x^2+1\right)\right]_0^\infty = +\infty.$ Therefore the given series diverges.

4. $\int_0^\infty (x+1)^{-4/3}\, dx = \left[-3\left(x+1\right)^{-1/3}\right]_0^\infty = 3 < \infty.$ Therefore the given series converges.

7. $\int_2^\infty \frac{1}{x\ln x}\, dx = \left[\ln\left(\ln x\right)\right]_2^\infty = +\infty.$

10. $\int_0^\infty xe^{-x}\, dx = 1 < \infty,$ so the given series converges.

13. $\int_1^\infty \frac{\ln x}{x^2}\, dx = \left[-\frac{1}{x}\left(1+\ln x\right)\right]_1^\infty = 1.$ (Integrate by parts; use l'Hôpital's rule to get the limit.)

16. $\int_1^\infty \frac{dx}{x^3+x} = \int_1^\infty \left(\frac{1}{x} - \frac{x}{x^2+1}\right)\, dx = \left[\left(\ln x\right) - \frac{1}{2}\ln\left(x^2+1\right)\right]_1^\infty = \frac{1}{2}\left[\ln\frac{x^2}{x^2+1}\right]_1^\infty = \frac{1}{2}\ln 2 < \infty,$
and therefore the given series converges.

19. Let $J = \int_1^\infty \ln\left(1+\frac{1}{x^2}\right)\, dx.$ Integrate by parts: With $u = \ln\left(1+x^{-2}\right)$ and $dv = dx$, we find that

$$J = \left[x\ln\left(1+x^{-2}\right)\right]_1^\infty + 2\int_1^\infty \frac{dx}{1+x^2}$$

$$= \lim_{x\to\infty} \frac{\ln\left(1+x^{-2}\right)}{1/x} - \ln 2 + \left[2\tan^{-1}x\right]_1^\infty$$

$$= \left(\lim_{u\to 0+} \frac{\ln\left(1+u^2\right)}{u}\right) + \pi - \frac{\pi}{2} - \ln 2 = \frac{\pi}{2} - \ln 2 < \infty,$$

so the series converges.

22. $\int_1^\infty \frac{x}{(4x^2+5)^{3/2}}\, dx = \left[-\frac{1}{4\sqrt{4x^2+5}}\right]_1^\infty < \infty.$ Therefore the given series converges.

28. $\int_1^\infty (x+1)^{-3}\, dx = \left[-\frac{1}{2}\left(x+1\right)^{-2}\right]_1^\infty = \frac{1}{8} < \infty.$ Therefore the given series converges.

34. The terms of the series are not monotonically decreasing. For example, the third term is about 4.9×10^{-6} but the fourth term is about 1.3×10^{-3}.

40. $n = 100.$

46. $n = 13;\ 1.0083493$

Section 11.6

1. Because $n^2 + n + 1 > n^2$ if $n > 0$, this series converges by comparison with the p-series for which $p = 2$.

4. Converges by limit-comparison with the p-series for which $p = \frac{3}{2}$.

7. Diverges by limit-comparison with the harmonic series.

10. Diverges by limit-comparison with the harmonic series.

13. Diverges by *careful* limit-comparison with the harmonic series.

16. Converges by comparison with the geometric series with ratio $\frac{1}{3}$.

19. Because $n^2 \ln n > n^2$ if $n \geq 3$, this series converges by comparison with the p-series for which $p = 2$.

22. Diverges by limit-comparison with the harmonic series.

25. Converges by comparison with the p-series for which $p = 2$. To show this, show that $\ln n < n$ if $n > 2$ and that $e^n > n^3$ if $n > 4$. To do the latter, sketch the graph of $f(x) = e^x - x^3$. Obtain an estimate of its largest x-intercept (it's approximately 4.5364) and show that the graph of $f(x)$ is increasing for all larger values of x.

28. Converges by comparison with the geometric series with ratio $\frac{1}{2}$.

34. Show that $\ln n < n$ if $n \geq 1$. Then the given series converges by comparison with the p-series for which $p = 2$.

40. The sum of the first ten terms (subscripts 2 through 11) is approximately 1.2248932892. The error is approximately 0.3947.

46. Let $f(x) = x - (\ln x)^8$. Then $f'(x) = 1 - \dfrac{8}{x}(\ln x)^7$. Now $f'(x) < 0$ when $8(\ln x)^7 < x$. By Newton's method we find that $8(\ln x)^7 = x$ when x is approximately 2.3104 and when x is approximately 4.18853×10^{10}. (In order to obtain the latter solution, apply Newton's method to the equation $\ln x = (x/8)^{1/7}$.) It is easy to show that $f'(x) > 0$ if $x > 4.2 \times 10^{10}$. Next, also by Newton's method, show that $f(x) = 0$ when x is about equal to 3.1752 and when x is approximately 2.1491×10^{11}. When all this information is assembled, you can see that

$$f(x) > 0 \quad \text{if} \quad x > 2.1492 \times 10^{11}.$$

Therefore $(\ln n)^8 < n$ for all $n > 2.1492 \times 10^{11}$, and so

$$\frac{1}{(\ln n)^8} > \frac{1}{n} \quad \text{if} \quad n > 2.1492 \times 10^{11}.$$

Section 11.7

1. This series meets both criteria of the alternating series test, and therefore it converges.

4. This series meets both criteria of the alternating series test, and therefore it converges.

7. Because $n/(\ln n) \to +\infty$ as $n \to \infty$, this series diverges by the nth-term test for divergence.

10. If a_n denotes the nth term of this series, then

$$|a_1| < |a_2| = |a_3| > |a_4| > |a_5| > |a_6| > |a_7| > \cdots,$$

and one can show that after the first two exceptions, the terms of this series steadily decreasing in magnitude and have limit zero. Changing the first two terms of a series cannot affect its convergence or divergence, so this series converges by the alternating series test.

13. This series has the same sequence of partial sums as the series

$$\sum_{n=1}^{\infty} \frac{(-1)^{n+1}}{(2n-1)^{2/3}} = 1 - \frac{1}{3^{2/3}} + \frac{1}{5^{2/3}} - \frac{1}{7^{2/3}} + \frac{1}{9^{2/3}} - \cdots,$$

and the latter series converges by the alternating series test. So the series given in Problem 13 also converges.

16. Because $n \sin(\pi/n) \to \pi \neq 0$ as $n \to \infty$, this series diverges by the n-th term test for divergence.

22. Diverges by limit-comparison with the harmonic series.

28. Converges absolutely by the ratio test (the limit in the ratio test is $\dfrac{1}{e}$, and $-1 < \dfrac{1}{e} < 1$).

34. Converges absolutely by the ratio test (the limit in the ratio test is $\frac{3}{4}$).

40. The ratio test yields the limit $\frac{2}{3}$, so the given series converges absolutely.

43. The sum of the first five terms of the series is approximately 0.904412037 and the sixth term is approximately -0.004629630. Hence the sum of the series lies between 0.899782406 and 0.904412038. We have only two-place accuracy; all we can say is that the sum of the series rounds to 0.90.

46. The sum of the first seven terms is approxiately 0.783430568 and the eighth term is approximately $-5.9604645 \times 10^{-8}$, so the sum of this series lies between 0.783430507 and 0.783430569, so we have six-place accuracy; the sum of the given series rounds to 0.783431.

49. The seventh term of the given series is approximately 0.000416493, so we should obtain three-place accuracy. The alternating series remainder estimate reveals that the sum of this series lies between 0.946767823 and 0.947184318, so to three places its sum is 0.947. The exact value of its sum is $\frac{7}{720}\pi^4$.

52. The sum of the first five terms (corresponding to $0 \leq n \leq 4$) of the series is approximately 0.540302579 and the sixth term (corresponding to $n = 5$) is approximately $-2.755731922 \times 10^{-7}$, so we have five-place accuracy; the sum of the series rounds to 0.54030.

58. This is merely the contrapositive of Theorem 3, so its proof is the same.

Section 11.8

1. The radius of convergence is

$$R = \lim_{n \to \infty} \left| \frac{n}{n+1} \right| = 1.$$

and the series clearly diverges when $x = \pm 1$. Hence the interval of convergence is $(-1, 1)$.

4. The radius of convergence is

$$R = \lim_{n \to \infty} \frac{5\sqrt{n+1}}{\sqrt{n}} = 5.$$

When $x = 5$, then the series converges by the alternating series test. When $x = -5$, it diverges because it dominates the harmonic series. Hence the interval of convergence is $(-5, 5]$.

7. The radius of convergence is

$$R = \lim_{n \to \infty} \frac{(n+1)^3}{3n^3} = \tfrac{1}{3}.$$

When $x = \tfrac{1}{3}$, the series converges because it is a p-series with $p = 3 > 1$. When $x = -\tfrac{1}{3}$, the series converges by the alternating series test. Hence the interval of convergence is $\left[-\tfrac{1}{3}, \tfrac{1}{3}\right]$.

10. The radius of convergence is $R = \lim_{n \to \infty} \dfrac{(3n+2)n^2}{(3n-1)(n+1)^2} = 1$. When $x = \pm 1$, the series diverges by the nth-term test for divergence, so its interval of convergence is $(-1, 1)$.

13. The radius of convergence is

$$R = \lim_{n \to \infty} \frac{3 \ln n}{\ln(n+1)} = 3.$$

When $x = \pm 3$, the series diverges by the nth-term test for divergence, so its interval of convergence is $(-3, 3)$.

16. $0 \leq x \leq 1$

22. $1 \leq x \leq 3$

28. $(-2, 2)$

31. Here we have

$$f(x) = \frac{x}{1-x} = x + x^2 + x^3 + x^4 + \cdots$$

with interval of convergence $(-1, 1)$.

34. One solution: By the method of partial fractions,

$$\frac{x}{9-x^2} = \frac{1}{2}\left(\frac{1}{3-x} - \frac{1}{3+x}\right),$$

which is the sum of two geometric series, each of which converges on $(-3,3)$. Hence their sum series

$$\sum_{n=1}^{\infty} \frac{x^{2n-1}}{3^{2n}} = \frac{x}{3^2} + \frac{x^3}{3^4} + \frac{x^5}{3^6} + \frac{x^7}{3^8} + \cdots$$

represents $f(x)$ on the interval $(-3,3)$.

40. $(9+x^3)^{-1/2} = \sum_{n=0}^{\infty} \frac{(-1)^n (2n)! \, x^{3n}}{(n!)^2 \, 2^{2n} 3^{2n+1}}$; the ratio test gives limit $\frac{1}{9}|x|^3$; the radius of convergence is therefore $9^{1/3}$.

43. Integrate $t^3 - \frac{1}{3!}t^9 + \frac{1}{5!}t^{15} - \frac{1}{7!}t^{21} + \cdots$.

46. $f(x) = x - \frac{x^3}{3^2} + \frac{x^5}{5^2} - \frac{x^7}{7^2} + \frac{x^9}{9^2} - \cdots$; the interval of convergence is $-1 \leq x \leq 1$.

49. Let $f(x) = \sum_{n=0}^{\infty} x^n$, $-1 < x < 1$. Then $xf'(x) = \frac{x}{(1-x)^2} = \sum_{n=1}^{\infty} nx^n$.

52. $2; \frac{3}{2}$

58. Multiply the series for $J_0(x)$ by x to obtain

$$xJ_0(x) = \sum_{n=0}^{\infty} \frac{(-1)^n x^{2n+1}}{2^{2n}(n!)^2}.$$

Termwise integration is "legal," and in this case, legal on the set of all real numbers. Hence

$$\int xJ_0(x)\, dx = \sum_{n=0}^{\infty} \frac{(-1)^n x^{2n+2}}{2 \cdot 2^{2n}(n+1)(n!)^2} = \sum_{n=0}^{\infty} \frac{(-1)^n x^{2n+2}}{2^{2n+1}n!(n+1)!} + C = J_1(x) + C.$$

64. We begin by replacing $\sin xt$ with its Maclaurin series:

$$\int_0^{\infty} e^{-t} \sin xt \, dt = \int_0^{\infty} e^{-t}\left(xt - \frac{x^3 t^3}{3!} + \frac{x^5 t^5}{5!} - \frac{x^7 t^7}{7!} + \cdots\right) dt$$

$$= x\int_0^{\infty} te^{-t}\, dt - \frac{x^3}{3!}\int_0^{\infty} t^3 e^{-t}\, dt + \frac{x^5}{5!}\int_0^{\infty} t^5 e^{-t}\, dt - \frac{x^7}{7!}\int_0^{\infty} t^7 e^{-t}\, dt + \cdots$$

$$= x - x^3 + x^5 - x^7 + \cdots = \frac{x}{1+x^2}$$

provided that $|x| < 1$.

Section 11.9

1. Use the binomial series for $4(x+1)^{1/3}$; take $x = \frac{1}{64}$.

4. We use the series

$$e^{-x} = 1 - \frac{1}{1!}x + \frac{1}{2!}x^2 - \frac{1}{3!}x^3 + \cdots.$$

To insure that $|R_n| < 0.0005$, we impose the condition that

$$\frac{(0.2)^{n+1}}{(n+1)!} < \frac{1}{2000},$$

and it suffices to take $n = 3$. We find that

$$e^{-0.2} \approx 1 - \frac{0.2}{1!} + \frac{(0.2)^2}{2!} - \frac{(0.2)^3}{3!} \approx 0.818666667.$$

Answer: 0.819. (The true value is approximately 0.818730753.)

10. First,

$$\cos x = 1 - \frac{1}{2!} x^2 + \frac{1}{4!} x^4 - \cdots ;$$

the error term in using the Taylor polynomial $P_n(x)$ to approximate $\cos x$ cannot exceed

$$\frac{|x|^{n+1}}{(n+1)!},$$

and here we have $x = \frac{\pi}{36} < 0.0873$. For error less than 0.0005, we need $n \geq 2$; we take $n = 2$, and the error will not exceed 0.000111. Result: $\cos(\pi/36) \approx 0.9961922823$. (The true value is approximately 0.9961946981.)

13. The Maclaurin series for the arctangent function yields

$$\frac{\arctan x}{x} = 1 - \tfrac{1}{3}x^2 + \tfrac{1}{5}x^4 - \tfrac{1}{7}x^6 + \cdots .$$

It follows that

$$\int_0^{1/2} \frac{\arctan x}{x}\, dx = \frac{1}{2} - \frac{1}{2^3 \cdot 3^2} + \frac{1}{2^5 \cdot 5^2} - \frac{1}{2^7 \cdot 7^2} + \cdots .$$

The least odd positive integer n for which

$$\frac{1}{2^n \cdot n^2} < 0.00005$$

is $n = 9$, so four terms of the last series yield

$$\int_0^{1/2} \frac{\arctan x}{x}\, dx \approx 0.4872,$$

which agrees, to four places, with the exact answer.

16. The binomial series yields

$$(1 + x^4)^{-1/2} = 1 - \frac{\tfrac{1}{2}}{1!} x^4 + \frac{\tfrac{1}{2} \cdot \tfrac{3}{2}}{2!} x^8 - \frac{\tfrac{1}{2} \cdot \tfrac{3}{2} \cdot \tfrac{5}{2}}{3!} x^{12}$$
$$+ \frac{3 \cdot 5 \cdot 7}{4! \cdot 2^4} x^{16} - \frac{3 \cdot 5 \cdot 7 \cdot 9}{5! \cdot 2^5} x^{20} + \cdots .$$

Then term-by-term integration yields

$$\int_0^{1/2} \frac{1}{\sqrt{1 + x^4}}\, dx = \frac{1}{2} - \frac{1}{2^6 \cdot 5} + \frac{3}{2! \cdot 2^{11} \cdot 9} - \frac{3 \cdot 5}{3! \cdot 2^{16} \cdot 13} + \frac{3 \cdot 5 \cdot 7}{4! \cdot 2^{21} \cdot 17} - \cdots .$$

Now

$$\frac{3}{2! \cdot 2^{11} \cdot 9} \approx 0.0008 \quad \text{but} \quad \frac{3 \cdot 5}{3! \cdot 2^{16} \cdot 13} \approx 0.000003,$$

so three terms of the last series yield the approximation

$$\int_0^{1/2} \frac{1}{\sqrt{1 + x^4}}\, dx \approx 0.4970.$$

19. The Maclaurin series for e^x yields

$$\exp(-x^2) = \frac{1}{0!} - \frac{x^2}{1!} + \frac{x^4}{2!} - \frac{x^6}{3!} + \cdots,$$

and then term-by-term integration gives

$$\int_0^1 \exp(-x^2)\, dx = 1 - \frac{1}{1! \cdot 3} + \frac{1}{2! \cdot 5} - \frac{1}{3! \cdot 7} + \cdots.$$

Next,

$$\frac{1}{6! \cdot 13} \approx 0.0001, \quad \text{and} \quad \frac{1}{7! \cdot 15} \approx 0.000013,$$

so six terms of the last series yield

$$\int_0^1 \exp(-x^2)\, dx \approx 0.7468.$$

22. First, the binomial series yields

$$(1 + x^3)^{-1/2} = 1 - \frac{1}{2}x^3 + \frac{3}{2! \cdot 2^2}x^6 - \frac{3 \cdot 5}{3! \cdot 2^3}x^9 + \frac{3 \cdot 5 \cdot 7}{4! \cdot 2^4}x^{12} - \cdots.$$

Then term-by-term integration gives

$$\int_0^{1/2} \frac{x}{\sqrt{1 + x^3}}\, dx = \frac{1}{8} - \frac{1}{320} + \frac{3}{16384} - \frac{5}{360448} + \frac{5}{4194304} - \cdots.$$

Now

$$\frac{5}{360448} \approx 0.000014, \quad \text{but} \quad \frac{5}{4194304} \approx 0.0000012,$$

So only four terms of the last series are need to give four-place accuracy:

$$\int_0^{1/2} \frac{x}{\sqrt{1 + x^3}}\, dx \approx 0.1220.$$

28. The limit can be simplified to $\lim\limits_{x \to 1} 2\left(1 - \frac{1}{2}(x-1) + \frac{1}{3}(x-1)^2 - \cdots\right) = 2.$

31. First we find that

$$\cos x = \frac{\sqrt{2}}{2}\left[\frac{1}{0!} - \frac{1}{1!}\left(x - \frac{\pi}{4}\right) - \frac{1}{2!}\left(x - \frac{\pi}{4}\right)^2 + \frac{1}{3!}\left(x - \frac{\pi}{4}\right)^3 + \frac{1}{4!}\left(x - \frac{\pi}{4}\right)^4 - \frac{1}{5!}\left(x - \frac{\pi}{5}\right)^5 - \cdots\right].$$

We convert $47°$ to radians and substitute to find that $\cos 47°$ is the sum of the series

$$\frac{\sqrt{2}}{2}\left[1 - \frac{\pi}{90} - \frac{1}{2!}\left(\frac{\pi}{90}\right)^2 + \frac{1}{3!}\left(\frac{\pi}{90}\right)^3 + \frac{1}{4!}\left(\frac{\pi}{90}\right)^4 - \frac{1}{5!}\left(\frac{\pi}{90}\right)^5 - \cdots\right].$$

We make this into an alternating series by grouping terms 2 and 3, terms 4 and 5, and so on. Let $x = \pi/90$. For six-place accuracy, we find that

$$\frac{x^3}{3!} + \frac{x^4}{4!} \approx 0.000007 \quad \text{and} \quad \frac{x^5}{5!} + \frac{x^6}{6!} \approx 0.00000000043,$$

so the first five terms of the (ungrouped) series—through exponent 4—yield the required six-place accuracy: $\cos 47° \approx 0.681998.$

34. The Taylor series remainder estimate yields

$$\frac{\cos z}{720}x^6 \leq \frac{10^{-6}}{720} \approx 1.4 \times 10^{-9} < 0.5 \times 10^{-8},$$

so eight-place accuracy is guaranteed (too conservative, as usual; you actually get 10-place accuracy).

37. Clearly $|e^z| < \frac{5}{3}$ if $|z| < 0.5$. Hence the Taylor series remainder estimate yields

$$\left| \frac{e^z}{120} x^5 \right| \le \frac{5}{3 \cdot 120} \left(\frac{1}{2} \right)^5 \approx 0.434 \times 10^{-3},$$

so three-place accuracy can be obtained if $|x| \le 0.5$. And

$$e^{1/3} \approx 1 + \frac{1}{3} + \frac{1}{18} + \frac{1}{486} + \frac{1}{1944} \approx 1.39.$$

55. A series approximation to the integrand is

$$1 - x^2 + x^6 - x^8 + x^{12} - x^{14} + \cdots + x^{60} - x^{62},$$

clearly far more terms than we need for two-place accuracy. Term-by-term integration of this series yields

$$\int_0^{1/2} \frac{1}{1 + x^2 + x^4} \, dx \approx 0.459239825,$$

and numerical integration with *Mathematica* 2.2.2 yields

```
NIntegrate[ 1/(1 + x^2 + x^4), {x, 0, 0.5} ]
0.459240
```

Chapter 11 Miscellaneous

1. $a_n = \dfrac{1 + (1/n^2)}{1 + (4/n^2)}$.

4. $a_n = 0$ for all n, so the limit is zero.

7. $-1 \le \sin x \le 1$ for all x, so $\{a_n\} \to 0$ by the squeeze theorem.

10. Apply l'Hôpital's rule thrice to obtain the limit zero.

13. Because e^n/n increases without bound and e^{-n}/n approaches zero, the limit does not exist (it is also correct to say that the given expression approaches $+\infty$).

16. The ratio test involves finding the limit of

$$\left(\frac{n}{n+1} \right)^n (n+1)$$

as n increases without bound. The first factor approaches $1/e$, so the limit is $+\infty$. Therefore the series diverges.

22. The given series diverges by the nth term test: $\lim\limits_{n \to \infty} -\dfrac{2}{n^2} = 0$, and therefore $\lim\limits_{n \to \infty} 2^{-(2/n^2)} = 1 \ne 0$.

28. $\lim\limits_{x \to \infty} x \sin \dfrac{1}{x} = \lim\limits_{u \to 0^+} \dfrac{\sin u}{u} = 1$, so the given series diverges by the nth -term test.

31. Is this not the Maclaurin series for $f(x) = e^{2x}$?

34. Limit: $\frac{1}{4} |2x - 3|$. Interval of convergence: $\left(-\frac{1}{2}, \frac{7}{2} \right)$.

40. Limit: $|x - 1|$. Interval of convergence: $(0, 2)$.

43. Use the ratio test.

46. If $\sum a_n$ converges, then $\{a_n\} \to 0$. So $a_n \le 1$ for all $n \ge N$ (where N is some sufficiently large integer). So the series $\sum a_n$ (eventually) dominates $\sum (a_n)^2$; therefore the latter series also converges.

52. $\ln(1+x) = x - \frac{1}{2}x^2 + \frac{1}{3}x^3 - \frac{1}{4}x^4 + \cdots$. For three-place accuracy, we need the first five terms of this series, and we find that

$$0.182266666 < \ln(1.2) < 0.182330667.$$

So $\ln(1.2) = 0.182$ to three places.

55. $\displaystyle\int_0^1 \left(1 - \frac{1}{2}x + \frac{1}{6}x^2 - \frac{1}{24}x^3 + \frac{1}{120}x^4 - \frac{1}{720}x^5 + \cdots\right) dx$

$$= \left[x - \frac{1}{4}x^2 + \frac{1}{18}x^3 - \frac{1}{96}x^4 + \frac{1}{600}x^5 - \frac{1}{4320}x^6 + \cdots\right]_0^1$$

$$1 - \frac{1}{2!\,2} + \frac{1}{3!\,3} - \frac{1}{4!\,4} + \frac{1}{5!\,5} - \frac{1}{6!\,6} + \cdots \approx 0.796599599297.$$

58. $\displaystyle\int_0^x \frac{dt}{1-t^2} = \int_0^x \left(1 + t^2 + t^4 + t^6 + \cdots\right) dt$

$$= \left[t + \frac{1}{3}t^3 + \frac{1}{5}t^5 + \frac{1}{7}t^7 + \cdots\right]_0^x$$

$$= x + \frac{1}{3}x^3 + \frac{1}{5}x^5 + \frac{1}{7}x^7 + \cdots = \sum_{n=0}^{\infty} \frac{x^{2n+1}}{2n+1}.$$

64. $\displaystyle P_k = \prod_{n=2}^{k} \frac{n^2}{n^2-1} = \left(\frac{2\cdot 2}{1\cdot 3}\right)\left(\frac{3\cdot 3}{2\cdot 4}\right)\left(\frac{4\cdot 4}{3\cdot 5}\right) \cdots \left(\frac{k\cdot k}{(k-1)\cdot(k+1)}\right) = \frac{2k}{k+1} \to 2$ as $k \to \infty$.

So $\displaystyle\prod_{n=2}^{\infty} \frac{n^2}{n^2-1} = 2$.

Chapter 12: Vectors, Curves, and Surfaces in Space

Section 12.1

1. $\mathbf{v} = \langle 3-1, 5-2 \rangle = \langle 2, 3 \rangle$

4. $\mathbf{v} = \langle 15 - (-10), -25 - 20 \rangle = \langle 35, -45 \rangle$

7. $(3\mathbf{i} + 5\mathbf{j}) + (2\mathbf{i} - 7\mathbf{j}) = 5\mathbf{i} + (-2)\mathbf{j} = 5\mathbf{i} - 2\mathbf{j}$

10. $|\mathbf{a}| = 5$, $|-2\mathbf{b}| = 10$, $|\mathbf{a} - \mathbf{b}| = 5\sqrt{2}$, $\mathbf{a} + \mathbf{b} = \langle -1, 7 \rangle$, and $3\mathbf{a} - 2\mathbf{b} = \langle 17, 6 \rangle$.

16. $\sqrt{2}$, $4\sqrt{2}$, $3\sqrt{2}$, $\mathbf{i} + \mathbf{j}$, and $-7\mathbf{i} - 7\mathbf{j}$.

22. $\mathbf{j}$

28. $\mathbf{a} \cdot \mathbf{b} = 120 - 120 = 0$; $\mathbf{a}$ and $\mathbf{b}$ are perpendicular.

31. $\langle 2, -3 \rangle = r\langle 1, 1 \rangle + s\langle 1, -1 \rangle$: $\quad r = -\frac{1}{2}$ and $s = \frac{5}{2}$.

34. (a) $12\mathbf{i} - 20\mathbf{j}$; (b) $\frac{3}{4}\mathbf{i} - \frac{5}{4}\mathbf{j}$

37. $(2c\mathbf{i} - 4\mathbf{j}) \cdot (3\mathbf{i} + c\mathbf{j}) = 0$ when $6c - 4c = 0$, thus when $c = 0$.

40. $(r + s)\langle a_1, a_2 \rangle = \langle (r+s)a_1, (r+s)a_2 \rangle$
$$= \langle ra_1 + sa_1, ra_2 + sa_2 \rangle$$
$$= \langle ra_1, ra_2 \rangle + \langle sa_1, sa_2 \rangle$$
$$= r\langle a_1, a_2 \rangle + s\langle a_1, a_2 \rangle = r\mathbf{a} + s\mathbf{a}.$$

43. With $\mathbf{T}_1 = \langle T_1 \cos 30°, T_1 \sin 30° \rangle$, $\mathbf{T}_2 = \langle -T_2 \cos 30°, T_2 \sin 30° \rangle$, and $\mathbf{F} = \langle 0, -100 \rangle$, the "equation of balance" $\mathbf{T}_1 + \mathbf{T}_2 + \mathbf{F} = 0$ yields $T_1 = T_2 = 100$.

46. With $\mathbf{T}_1 = \langle \frac{3}{5}T_1, \frac{4}{5}T_1 \rangle$, $\mathbf{T}_2 = \langle -\frac{4}{5}T_2, \frac{3}{5}T_2 \rangle$, and $\mathbf{F} = \langle 0, -150 \rangle$, the balance equation $\mathbf{T}_1 + \mathbf{T}_2 + \mathbf{F} = 0$ yields $T_1 = 120$ and $T_2 = 90$.

52. $\overrightarrow{AM_1} + \overrightarrow{M_1B} = \overrightarrow{AB}$, $\overrightarrow{CM_2} + \overrightarrow{M_2A} = \overrightarrow{CA}$, and $\overrightarrow{AB} + \overrightarrow{BC} + \overrightarrow{CA} = 0$.
But $\overrightarrow{AM_1} = \overrightarrow{M_1B}$ and $\overrightarrow{AM_2} = \overrightarrow{M_2C}$. So $2\overrightarrow{AM_1} + 2\overrightarrow{M_2A} + \overrightarrow{BC} = 0$.
Also $\overrightarrow{AM_1} + \overrightarrow{M_2A} + \overrightarrow{M_1M_2} = 0$. Therefore $-2\overrightarrow{M_1M_2} + \overrightarrow{BC} = 0$; that is, $\overrightarrow{M_1M_2} = \frac{1}{2}\overrightarrow{BC}$.

Section 12.2

4. $9\mathbf{i} - 3\mathbf{j} + 3\mathbf{k}$; $-14\mathbf{i} - 21\mathbf{j} + 43\mathbf{k}$; -34; $3\sqrt{21}$; $(2\mathbf{i} - 3\mathbf{j} + 5\mathbf{k})/\sqrt{38}$

7. $80°$

10. $127°$

16. $\text{comp}_{\mathbf{b}}\mathbf{a} = -\dfrac{34}{\sqrt{83}}$; $\text{comp}_{\mathbf{a}}\mathbf{b} = -\dfrac{34}{\sqrt{38}}$.

19. $x^2 + y^2 + z^2 - 6x - 2y - 4z = 11$

22. $x^2 + y^2 + z^2 - 8x - 10y + 4z + 7 = 0$

28. $x^2 + y^2 + z^2 - \frac{7}{2}x - \frac{9}{2}y - \frac{11}{2}z = 0$: $\left(x - \frac{7}{4}\right)^2 + \left(y - \frac{9}{4}\right)^2 + \left(z - \frac{11}{4}\right)^2 = \frac{49}{16} + \frac{81}{16} + \frac{121}{16} = \frac{251}{16}$; center $\left(\frac{7}{4}, \frac{9}{4}, \frac{11}{4}\right)$, radius $\frac{1}{4}\sqrt{251}$.

34. $x^2 + y^2 + z^2 + 7 = 0$: No points satisfy this equation.

37. $(x - 3)^2 + (y + 4)^2 + (z - 0)^2 = 0$: The single point $(3, -4, 0)$.

40. Each of $\mathbf{b} = r\mathbf{a}$ and $\mathbf{a} = s\mathbf{b}$ leads to a contradiction, so $\mathbf{a}$ and $\mathbf{b}$ are not parallel; $\mathbf{a} \cdot \mathbf{b} = 24$, so $\mathbf{a}$ and $\mathbf{b}$ are not perpendicular.

43. $\overrightarrow{QR} = 3\overrightarrow{PQ}$, so P, Q, and R are collinear.

46. $\angle A \approx 56°$, $\angle B = 90°$, $\angle C \approx 34°$

52. $\alpha \approx 74.2188°$, $\beta \approx 49.2535°$, $\gamma = 45°$

55. $40(\cos 40°) \cdot 1000 \cdot (0.239) \approx 7323.385$ (cal) (less than 8 Cal)

58. If θ is the angle between $\mathbf{a}$ and $\mathbf{b}$, then $\cos\theta = \dfrac{\mathbf{a} \cdot \mathbf{b}}{|\mathbf{a}| \, |\mathbf{b}|}$. Therefore, because $-1 \leq \cos\theta \leq 1$,

$\dfrac{\mathbf{a} \cdot \mathbf{b}}{|\mathbf{a}| \, |\mathbf{b}|} \leq 1$. Consequently $\mathbf{a} \cdot \mathbf{b} \leq |\mathbf{a}| \, |\mathbf{b}|$.

61. The simultaneous equations $\mathbf{u} \cdot \mathbf{w} = 0 = \mathbf{v} \cdot \mathbf{w}$ yield the scalar equations

$$w_1 + 2w_2 - 3w_3 = 0,$$
$$2w_1 + w_3 = 0,$$

which in turn yield $w_2 = -\frac{7}{2}w_1$ and $w_3 = -2w_1$ where w_1 is an arbitrary nonzero scalar. For example, $\mathbf{w} = \langle 2, -7, -4 \rangle$.

70. If $A = (0,0,0)$, $B = (1,1,0)$, and $C = (\frac{1}{2}, \frac{1}{2}, \frac{1}{2})$, then the angle between $\overrightarrow{CA}$ and $\overrightarrow{CB}$ is approximately 1.910633236; that is, about $109° \, 28' \, 16.4''$.

Section 12.3

1. $\mathbf{a} \times \mathbf{b} = \begin{vmatrix} \mathbf{i} & \mathbf{j} & \mathbf{k} \\ 5 & -1 & -2 \\ -3 & 2 & 4 \end{vmatrix} = (-4+4)\mathbf{i} - (20-6)\mathbf{j} + (10-3)\mathbf{k}.$

4. $-24\mathbf{j} - 24\mathbf{k}$

7. $\mathbf{c} = \mathbf{a} \times \mathbf{b} = \langle 36, -9, 12 \rangle$, so $\mathbf{v} = -\mathbf{u} = \dfrac{\mathbf{c}}{|\mathbf{c}|} = \frac{1}{13}\langle 12, -3, 4 \rangle.$

16. $\overrightarrow{OP} = \langle 1, 1, 0 \rangle$, $\overrightarrow{OQ} = \langle 1, 0, 1 \rangle$, and $\overrightarrow{OR} = \langle 0, 1, 1 \rangle$. Hence the volume V of the parallelepiped is given by

$$V = |\overrightarrow{OP} \times \overrightarrow{OQ} \cdot \overrightarrow{OR}| = \text{ABS} \begin{vmatrix} 1 & 1 & 0 \\ 1 & 0 & 1 \\ 0 & 1 & 1 \end{vmatrix} = 2.$$

19. $(\overrightarrow{AB} \times \overrightarrow{AC}) \cdot \overrightarrow{AD} = 0$, so the four given points are coplanar.

22. $(\overrightarrow{AB} \times \overrightarrow{AC}) \cdot \overrightarrow{AD} = 54 \neq 0$, so the four given points are not coplanar. The volume of the pyramid is 9.

34. First, $(\mathbf{a} \times \mathbf{b}) \times (\mathbf{c} \times \mathbf{d})$ is perpendicular to $\mathbf{a} \times \mathbf{b}$ and $\mathbf{a} \times \mathbf{b}$ is perpendicular to $\mathbf{a}$ and to $\mathbf{b}$. So $(\mathbf{a} \times \mathbf{b}) \times (\mathbf{c} \times \mathbf{d})$ lies in the plane determined by $\mathbf{a}$ and $\mathbf{b}$; that is,

$$(\mathbf{a} \times \mathbf{b}) \times (\mathbf{c} \times \mathbf{d}) = r_1\mathbf{a} + r_2\mathbf{b}.$$

(If $\mathbf{a}$ is parallel to $\mathbf{b}$, then $\mathbf{a} \times \mathbf{b} = 0$; in this case, take $r_1 = 0 = r_2$.) The other case follows by symmetry (or by interchanging $\mathbf{a}$ with $\mathbf{c}$ and $\mathbf{b}$ with $\mathbf{d}$ in the above proof).

Section 12.4

4. $x = 17 - 17t$, $y = 13t - 13$, $z = 31t - 13$

7. $\langle x, y, z \rangle = \langle 3, 5, 7 \rangle + t \langle 3, 0, -3 \rangle$ yields $x = 3t + 3$, $y \equiv 5$, $z = 7 - 3t$.

10. $x = 2 + 2t$, $y = 5 - 2t$, $z = -7 + 15t$; $\quad \dfrac{x-2}{2} = \dfrac{y-5}{-2} = \dfrac{z+7}{15}.$

16. A vector parallel to L_1 is $\mathbf{v}_1 = \langle 4, 1, -2 \rangle$ and a vector parallel to L_2 is $\mathbf{v}_2 = \langle 6, -3, 8 \rangle$. These vectors are not parallel because $\mathbf{v}_1 \times \mathbf{v}_2 = \langle 2, -44, -18 \rangle \neq \mathbf{0}$. Hence L_1 and L_2 are not parallel.

When we solve their equations simultaneously, we find the single solution $(x, y, z) = (7, 5, -3)$, so the two lines meet in a single point.

19. A vector parallel to L_1 is $\mathbf{v}_1 = \langle 6, 4, -8 \rangle$ and a vector parallel to L_2 is $\mathbf{v}_2 = \langle 9, 6, -12 \rangle = \frac{3}{2}\mathbf{v}_1$, so L_1 and L_2 are parallel. They do not intersect because $(7, -5, 9)$ lies on L_1 but not on L_2.

22. $2x - 7y - 3z = 19$

28. $x - 3y + 2z = 14$

34. Two points on L are $(17, 23, 35)$ and $(8, 37, -6)$, so two vectors parallel to the plane are $\mathbf{a} = \langle 4, \, 30, \, 6 \rangle$ and $\mathbf{b} = \langle -5, \, 44, \, -35 \rangle$. Hence a normal to that plane is $\mathbf{n} = \mathbf{a} \times \mathbf{b} = \langle -1314, \, 110, \, 326 \rangle$. It follows that an equation of the plane is $657x - 55y - 163z = 4199$.

37. The four given equations have the unique simultaneous solutions $x = \frac{9}{2}$, $y = \frac{9}{4}$, $z = \frac{17}{4}$, $t = \frac{3}{4}$, so the line and the plane are not parallel and meet at the point $\left(\frac{9}{2}, \frac{9}{4}, \frac{17}{4} \right)$.

40. Normals: $\langle 2, -1, 1 \rangle$ and $\langle 1, 1, -1 \rangle$. If θ is the angle between the two planes then it follows immediately that $\cos\theta = 0$, and therefore the angle between the two planes is $\pi/2$.

43. The cross product of the obvious two normals to the two planes is $\mathbf{v} = \langle 0, -1, 1 \rangle$. So the line of intersection has vector equation

$$\langle x, \, y, \, z \rangle = \langle 10, \, 0, \, 0 \rangle + t\,\mathbf{v},$$

and thus parametric equations $x = 10$, $y = -t$, $z = t$. Technically, this line doesn't have symmetric equations, but elimination of t yields its Cartesian equations

$$x = 10, \qquad y + z = 0.$$

46. If the equation $z = 0$ is adjoined to the equations of the two planes, the resulting system has simultaneous solution $(x, y, z) = \left(\frac{7}{5}, \frac{6}{5}, 0 \right)$, so this is one point on the line of intersection. Similarly, using $z = 1$ in place of $z = 0$ yields a second point of intersection, $\left(\frac{7}{5}, \frac{1}{5}, 1 \right)$. Hence the line is parallel to $\mathbf{v} = \langle 0, -1, 1 \rangle$. It has vector equation $\langle x, \, y, \, z \rangle = \frac{1}{5}\langle 7, 6, 0 \rangle + t\,\mathbf{v}$ and thus parametric equations

$$x \equiv \tfrac{7}{5}, \quad y = \tfrac{6}{5} - t, \quad z = t.$$

Elimination of the parameter yields the Cartesian equations

$$x = \tfrac{7}{5}, \quad y = \tfrac{6}{5} - z.$$

49. The xy-plane has equation $z = 0$, so the plane with equation $3x + 2y - z = 6$ meets it in the line with equation $3x + 2y = 6$. Two points on this line are $P(2, 0, 0)$ and $Q(0, 3, 0)$. In order to contain the third point $R(1, 1, 1)$, the plane in question must have normal vector $\overrightarrow{PQ} \times \overrightarrow{PR} = \langle 3, 2, 1 \rangle$. Hence it has an equation of the form $3x + 2y + z = D$ for some constant D. Because the point $(1, 1, 1)$ is in the plane, $D = 6$.

52. Suppose that the lines meet at (x, y, z). Then the simultaneous equations that result have solution $P(1, -1, 2)$ The first line also contains $Q(2, 1, 3)$ and the second also contains $R(3, 5, 6)$. A normal to the plane is $\overrightarrow{PQ} \times \overrightarrow{PR} = \langle 2, -2, 2 \rangle$; use $\langle 1, -1, 1 \rangle$ instead. Answer: $x - y + z = 4$.

58. You may assume without loss of generality that $a \neq 0$. Then the point $(-d_1/a, 0, 0)$ lies on the first plane. The desired result now follows.

61. A vector parallel to L is $\mathbf{v} = \langle 3, -9, 1 \rangle$. Three points in the plane $\mathcal{P}$ are $P = (0, 0, 0)$, $Q = (5, 7, 3)$, and $R = (4, -1, 2)$. A normal to $\mathcal{P}$ is $\mathbf{n} = \overrightarrow{PQ} \times \overrightarrow{PR} = \langle 17, 2, -33 \rangle$. The angle between $\mathbf{n}$ and $\mathbf{v}$ is

$$\theta = \arccos\left(\frac{\mathbf{n} \cdot \mathbf{v}}{|\mathbf{n}||\mathbf{v}|} \right) = \frac{\pi}{2},$$

so L and $\mathcal{P}$ are parallel.

64. A vector parallel to L is $\mathbf{v} = \langle -6, 3, 3 \rangle$. Three points in $\mathcal{P}$ are $P = (0,0,0)$, $Q = (7, 4, -5)$, and $R = (3, -2, 6)$. So a normal to $\mathcal{P}$ is $\mathbf{n} = \overrightarrow{PR} \times \overrightarrow{PQ} = \langle -14, 57, 26 \rangle$. The angle between $\mathbf{v}$ and $\mathbf{n}$ is

$$\theta = \arccos \left(\frac{\mathbf{n} \cdot \mathbf{v}}{|\mathbf{n}||\mathbf{v}|} \right) = \arccos \left(\frac{111}{\sqrt{24726}} \right) \approx 0.7870967836.$$

We convert to degrees and subtract from $90°$ to obtain the angle between L and $\mathcal{P}$; it is approximately $44.9027°$. Finally, we solve the equations of L and $\mathcal{P}$ simultaneously to find that they meet at the point $(13, 0, 7)$.

67. Three points on $\mathcal{P}_1$ are $P_1 = (0,0,0)$, $Q_1 = (7, 8, 7)$, and $R_1 = (1, 12, 5)$. So a normal to $\mathcal{P}_1$ is

$$\mathbf{n}_1 = \overrightarrow{P_1 R_1} \times \overrightarrow{P_1 Q_1} = \langle 44, -28, -76 \rangle.$$

Three points on $\mathcal{P}_2$ are $P_2 = (0,0,0)$, $Q_2 = (-3, 7, -2)$, and $R_2 = (4, 4, 11)$. So a normal to $\mathcal{P}_2$ is

$$\mathbf{n}_2 = \overrightarrow{P_2 Q_2} \times \overrightarrow{P_2 R_2} = \langle 85, 25, -40 \rangle = 7\mathbf{n}_1.$$

The angle between the planes is the angle between their normals,

$$\theta = \arccos \left(\frac{\mathbf{n}_1 \cdot \mathbf{n}_2}{|\mathbf{n}_1||\mathbf{n}_2|} \right) = \arccos \left(\frac{374}{\sqrt{200718}} \right) \approx 0.5830408732.$$

In degrees, the angle is approximately $33.4057813183°$.

The line of intersection of $\mathcal{P}_1$ and $\mathcal{P}_2$ is parallel to

$$\mathbf{v} = \tfrac{1}{20} (\mathbf{n}_1 \times \mathbf{n}_2) = \langle -39, 235, 64 \rangle.$$

One way to proceed is to find a point at which $\mathcal{P}_1$ intersects $\mathcal{P}_2$. We solved the system

$$x = 7s + t = -3u + 4v,$$
$$y = 8s + 12t = 7u + 4v,$$
$$z = 7s + 5t = -2u + 11v$$

to find

$$x = -\tfrac{65}{18}v, \qquad y = \tfrac{1175}{54}v, \qquad z = \tfrac{160}{27}v.$$

Then the choice $v = 54$ gives a point of intersection with integral coordinates, $P = (-195, 1175, 320)$. A vector equation of the line of intersection is then

$$\langle x, y, z \rangle = \langle -195, 1175, 320 \rangle + t\,\mathbf{v},$$

and so parametric equations of this line are

$$x = -39t - 195, \quad y = 235t + 1175, \quad z = 64t + 320.$$

Remember that parametric equations of lines are not unique. A simple substitution allows you to present the parametric equations of this line in the simpler form

$$x = -39t, \quad y = 235t, \quad z = 64t.$$

Section 12.5

1. Note that $y^2 + z^2 \equiv 1$. The steady change in x implies that this curve is the helix of Fig. 12.5.17.

4. It's easy to show that $x^2 + y^2 + z^2 \equiv 1$. Hence this must be the curve shown in Fig. 12.5.15.

10. $\mathbf{r}'(t) = \langle -5\sin t, 4\cos t \rangle$, $\mathbf{r}''(t) = \langle -5\cos t, -4\sin t \rangle$. So $\mathbf{r}'(\pi) = \langle 0, -4 \rangle$ and $\mathbf{r}''(\pi) = \langle 5, 0 \rangle$.

16. $\mathbf{v}(t) = \langle 12, 10\cos 2t, 10\sin 2t \rangle$; $\mathbf{a}(t) = \langle 0, -20\sin 2t, 20\cos 2t \rangle$; $v(t) = \sqrt{244}$

22. $\langle 1, 2t \rangle \cdot \langle t^2, -t \rangle + \langle t, t^2 \rangle \cdot \langle 2t, -1 \rangle = t^2 - 2t^2 + 2t^2 - t^2 = 0$.

25. $\mathbf{v}(t) = \langle 0, 0, 1 \rangle$ and $\mathbf{r}(t) = \langle 1, 0, t \rangle$.

28. $\mathbf{v}(t) = \langle t, 7 - t, 3t \rangle$ and $\mathbf{r}(t) = \langle 5 + \frac{1}{2}t^2, 7t - \frac{1}{2}t^2, \frac{3}{2}t^2 \rangle$.

34. $\mathbf{v}(t) = \langle -1 + 3\cos 3t, -3\sin 3t, 4t - 7 \rangle$ and $\mathbf{r}(t) = \langle 3 - t - \sin 3t, 3 + \cos 3t, 2t^2 - 7t \rangle$.

37. Given $\mathbf{u}(t) = \langle 0, 3, 4t \rangle$ and $\mathbf{v}(t) = \langle 5t, 0, -4 \rangle$, we have

$$\mathbf{u}(t) \times \mathbf{v}(t) = \langle -12, 20t^2, -15t \rangle,$$

with derivative

$$\langle 0, 40t, -15 \rangle = \langle 0, 20t, -15 \rangle + \langle 0, 20t, 0 \rangle$$
$$= \langle 0, 3, 4t \rangle \times \langle 5, 0, 0 \rangle + \langle 0, 0, 4 \rangle \times \langle 5t, 0, -4 \rangle$$
$$= \mathbf{u}(t) \times \mathbf{v}'(t) + \mathbf{u}'(t) \times \mathbf{v}(t).$$

40. Let $\mathbf{v}(t)$ denote the velocity vector of the moving particle. If the speed $|\mathbf{v}(t)|$ of the particle is constant, then $\mathbf{v}(t) \cdot \mathbf{v}(t) = C$ where C is a constant. Now differentiate each side of this identity with respect to time.

43. Suppose that a projectile is launched from the point (x_0, y_0), with y_0 denoting its initial height above the surface of the earth. Let α be the angle of inclination from the horizontal of its initial velocity vector $\mathbf{v}_0$, with $\alpha > 0$ if the initial velocity vector points generally upward. Suppose also that the motion of the projectile takes place sufficiently close to the surface of the earth that we may assume the acceleration g of gravity to be constant. If we also ignore air resistance, then we may derive the equations of motion of the projectile:

$$x = (v_0 \cos \alpha)t + x_0, \tag{1}$$
$$y = -\tfrac{1}{2}gt^2 + (v_0 \sin \alpha)t + y_0. \tag{2}$$

We will use these in several of the next few solutions. To solve Problem 43, first derive the formula
$$(v_0)^2 = \frac{Rg}{\sin 2\alpha}.$$

46. As a consequence of either Problem 22 or Problem 23,

$$R = \frac{1}{g}(v_0)^2 \sin 2\alpha, \quad \text{so}$$

$$\frac{dR}{d\alpha} = \frac{2}{g}(v_0)^2 \cos 2\alpha.$$

Now $dR/d\alpha = 0$ when $\cos 2\alpha = 0$, so that $\alpha = \pi/4$ is the angle that maximizes the range R.

49. See the diagram on the next page.

$$\mathbf{a} = -(9.8)\mathbf{j},$$
$$\mathbf{v}(t) = v_0\mathbf{i} - (9.8)t\,\mathbf{j}, \quad \text{and}$$
$$\mathbf{r}(t) = v_0 t\,\mathbf{i} + \left(100 - (4.9)t^2\right)\mathbf{j}.$$

At the time T of impact,

$$v_0 T = 1000 \quad \text{and} \quad 100 - (4.9)T^2 = 0.$$

We solve for T: $T = \frac{10}{7}\sqrt{10}$; so

$$v_0 = \frac{7000}{10\sqrt{10}} \approx 221.36 \ \text{m/s}.$$

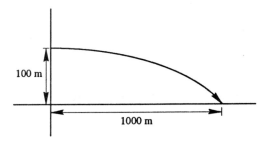

100 m

1000 m

52. The equations of motion of the projectile are

$$x(t) = (100 \cos \alpha)\, t \quad \text{and} \quad y(t) = -16t^2 + (1000 \sin \alpha)\, t + 500$$

if we take the gun at position $(0, 500)$ in the usual coordinate system. We want $x(T) = 20,000$ when $y(T) = 0$. The first of these conditions implies that $T = 20 \sec \alpha$, and the second yields

$$-16T^2 + (1000 \sin \alpha)\, T + 500 = 0.$$

Substitution for T in the last equation gives

$$64u^2 - 200u + 59 = 0 \quad \text{where} \quad u = \tan T.$$

This quadratic equation has two solutions, yielding

$$u \approx 2.795193, \quad \alpha \approx 70.314970°, \quad T \approx 59.3737 \ (\text{sec}) \quad \text{and}$$
$$u \approx 0.329807, \quad \alpha \approx 18.252934°, \quad T \approx 21.0597 \ (\text{sec}).$$

Either initial angle will "work", but the second is likely to be much more accurate.

58. $\mathbf{r}(t) = \langle a \cos \omega t, b \sin \omega t \rangle$;

$\mathbf{v}(t) = \langle -a\omega \sin \omega t, b\omega \cos \omega t \rangle$,

$\mathbf{a}(t) = \langle -a\omega^2 \cos \omega t, -b\omega^2 \sin \omega t \rangle = -\omega^2 \langle a \cos \omega t, b \sin \omega t \rangle$.

Therefore $\mathbf{a}(t) = c\mathbf{r}(t)$ where $c = -\omega^2 < 0$. So $\mathbf{F}(t)$ is a force directed toward the origin and has magnitude proportional to distance from the origin.

64. We assume that the baseball is hit directly down the left-field foul line and that this is the direction of due north. We assume also that the acceleration due to spin is directed due east. Then the acceleration vector of the baseball is

$$\mathbf{a}(t) = \langle 2, 0, -32 \rangle.$$

The initial velocity of the baseball is $\mathbf{v}_0 = \langle 0, 96 \cos 15°, 96 \sin 15° \rangle$, so its velocity vector is

$$\mathbf{v}(t) = \left\langle 2t, \ 24 \left(1 + \sqrt{3}\right) \sqrt{2}, \ 24 \left(-1 + \sqrt{3}\right) \sqrt{2} - 32t \right\rangle.$$

The initial position of the baseball is $\mathbf{r}_0 = \langle 0, 0, 0 \rangle$ in our coordinate system, so the baseball has position vector

$$\mathbf{r}(t) = \left\langle t^2, \ 24t \left(1 + \sqrt{3}\right) \sqrt{2}, \ 24t \left(-1 + \sqrt{3}\right) \sqrt{2} - 16t^2 \right\rangle.$$

The ball strikes the ground when the third component of $\mathbf{r}(t)$ is zero; that is, when

$$t = \tfrac{3}{2} \left(\sqrt{6} - \sqrt{2}\right).$$

At this time the x-component of $\mathbf{r}(t)$ is $18 - 9\sqrt{3} \approx 2.411543$, so the ball hits the ground just under 2 ft 5 in. from the foul line.

Section 12.6

1. $\displaystyle\int_0^\pi \sqrt{(6\cos 2t)^2 + (6\sin 2t)^2 + 64}\ dt = 10\pi.$

4. $(ds/dt)^2 = t^2 + (1/t)^2 + 2 = (t + t^{-1})^2$; integrate ds from $t = 1$ to $t = 2$ to get $s = \frac{3}{2} + \ln 2.$

10. The curvature is $\dfrac{|(1)\,(2) - (0)\,(7)|}{(1^2 + 7^2)^{3/2}} = \dfrac{1}{250}\sqrt{2}.$

16. The curvature at x is $f(x) = \dfrac{2\,|x|^3}{(x^4 + 1)^{3/2}}$. It suffices to consider the case in which $x > 0$; if so, then

$$f'(x) = \frac{6x^2\,(1 - x^4)}{(x^4 + 1)^{5/2}}.$$

It follows that $x = 1$ maximizes $f(x)$, its maximum value is $1/\sqrt{2}$, and the points on the graph of the equation $xy = 1$ where the curvature is maximal are $(1, 1)$ and $(-1, -1)$.

22. $\mathbf{a}(t) = \langle -3\pi^2 \sin \pi t, -3\pi^2 \cos \pi t\rangle$. So $a_T = 0$, $a_N = 3\pi^2.$

34. $\mathbf{v} = \langle 1, 2t, 3t^2\rangle$; $\mathbf{a} = \langle 0, 2, 6t\rangle$. So $\mathbf{v} \times \mathbf{a} = \langle 6t^2, -6t, 2\rangle$. $v = \sqrt{1 + 4t^2 + 9t^4}$;

$|\mathbf{v} \times \mathbf{a}| = \sqrt{36t^4 + 36t^2 + 4}$. Thus the curvature is $\dfrac{2\sqrt{9t^4 + 9t^2 + 1}}{(9t^4 + 4t^2 + 1)^{3/2}}.$

40. $|\mathbf{v} \times \mathbf{a}| = \sqrt{6}e^{2t}$, $v = |\mathbf{v}| = \sqrt{3}e^t$, and the curvature is $\frac{1}{3}\sqrt{2}e^{-t}.$

Therefore $\dfrac{dv}{dt} = a_T = \sqrt{3}e^t$ and $a_N = \sqrt{2}e^t.$

46. The curvature is $\dfrac{a\omega^2}{a^2\omega^2 + b^2}$. Also $\dfrac{dv}{dt} = 0$, so $a_T = 0$. Next, $a_N = \kappa v^2 = a\omega^2$. It follows that

$$\mathbf{T}(t) = \frac{\langle -a\omega \sin \omega t, a\omega \cos \omega t, b\rangle}{\sqrt{a^2\omega^2 + b^2}}; \quad \mathbf{N}(t) = -\langle \cos \omega t, \sin \omega t, 0\rangle; \quad a_T = 0; \quad a_N = a\omega^2.$$

49. $s = s(t) = \displaystyle\int_0^t ds = \int_0^t \sqrt{9\sin^2 t + 9\cos^2 t + 16}\ dt = 5t.$ Therefore $t = s/5$ and the rest is easy.

58. With the usual meanings of the symbols, we have $b^2 = a^2\,(1 - e^2)$. With the aid of a calculator and the result of Problem 8, we obtain these results: Its velocity at apogee is approximately 18.2007 miles per second; at perigee, it is approximately 18.8189 miles per second.

61. Equation (11), when applied to the moon, yields $(26.32)^2 = k\,(238,850)^3.$

For a satellite with period $T = \frac{1}{24}$ (of a day—one hour), it yields $1^2 = kr^3$ where r is the radius of the orbit of the satellite. Divide the second of these equations by the first and solve for $r \approx 3165$ miles, about 795 miles below the earth's surface. So it can't be done.

Section 12.7

1. A plane with intercepts $x = \frac{20}{3}$, $y = 10$, $z = 2.$

4. A cylinder whose rulings are lines parallel to the z-axis. It meets the xy-plane in the hyperbola $x^2 - y^2 = 9$. The graph is shown at the right.

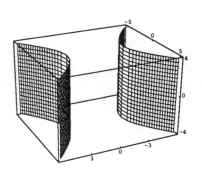

7. An elliptic paraboloid. The graph is shown on the right.

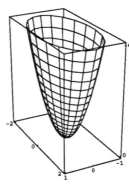

10. A circular cylinder with axis the x-axis.

16. Parabolic cylinder, parallel to the z-axis; its trace in the xz-plane is the parabola with vertex $x = 9$ and $z = 0$ and opening in the negative x-direction. The graph is shown at the right.

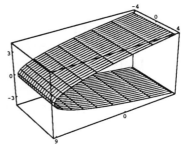

22. A cylinder having the graph $x = \sin y$ as its trace in the xy-plane. The graph is shown below.

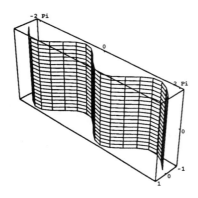

28. A hyperboloid of one sheet. Its graph is shown on the right.

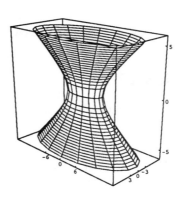

34. Replace x by $\sqrt{x^2 + y^2}$. Answer: A paraboloid opening along the negative z-axis and with equation $z = 4 - x^2 - y^2$.

40. $4x^2 = y^2 + z^2$

46. Hyperbolas (both branches), except that the trace in the xy-plane itself consists of the coordinate axes.

52. Projection of the intersection: $2x^2 + 3z^2 = 5 - 3x^2 - 2z^2$. That is, $x^2 + z^2 = 1$: a circle.

58. $\lambda_1 = -5$, $\lambda_2 = 10$: hyperbola

64. $\lambda_1 = 5$, $\lambda_2 = 10$: elliptic paraboloid

70. $\lambda_1 \approx 0.492097$, $\lambda_2 \approx 3.24436$, $\lambda_3 \approx 6.26354$: ellipsoid

Section 12.8

4. $\left(-\frac{3}{2}\sqrt{3}, -\frac{3}{2}, -1\right)$

10. $\left(-1, \sqrt{3}, 2\sqrt{3}\right)$

13. Cylindrical: $(0, \pi/2, 5)$; spherical: $(5, 0, \pi/2)$

16. Cylindrical: $\left(2\sqrt{2}, -\pi/4, 0\right)$; spherical: $\left(2\sqrt{2}, \pi/2, -\pi/4\right)$

22. Cylindrical: $\left(2\sqrt{5}, \tan^{-1}(-2), -12\right)$; spherical: $\left(2\sqrt{41}, \cos^{-1}\left(-\sqrt{41}/6\right), \tan^{-1}(-2)\right)$

28. The lower nappe of the cone $z^2 = 3r^2$, vertex at the origin, axis the z-axis, opening downward.

34. $x^2 + y^2 + (z - 2)^2 = 4$: This is the spherical surface with center $(0, 0, 2)$ and radius 2.

40. Cylindrical equation: $r = 2\cos\theta$. Spherical equation: $\rho\sin\phi = 2\cos\theta$.

46. The graph is a hemispherical surface: the upper half of the spherical surface with center $(0, 0, 0)$ and radius 2. To see a graph, execute the *Mathematica* command

```
ParametricPlot3D[ {2*Sin[phi]*Cos[theta], 2*Sin[phi]*Sin[theta],
    2*Cos[phi]}, {phi, 0, Pi/2}, {theta, 0, 2*Pi} ];
```

52. This solid region is bounded above by the sphere of radius 10 centered at the origin and bounded below by the 30°-cone, opening upward, with vertex at the origin. To see a figure, execute the *Mathematica* commands

```
p1 = ParametricPlot3D[ {10*Sin[phi]*Cos[theta], 10*Sin[phi]*Sin[theta],
    10*Cos[phi]}, {phi, 0, Pi/6}, {theta, 0, 2*Pi} ];
p2 = ParametricPlot3D[ {r*Cos[theta], r*Sin[theta], r*Sqrt[3]},
    {r, 0, 5}, {theta, 0, 2*Pi} ];
Show[ p1, p2 ]
```

58. The 62nd parallel is at (approximately) $z = 3496.47$; for such points we have the cylindrical equation

$$r = \sqrt{R^2 - z^2} \approx 1859.1$$

(the units are all miles). The angular difference in longitude is $178.28°$, so the distance along the parallel is given by $s = r\theta$ where θ is the radian measure of $178.28°$. Answer: About 5785 miles.

Chapter 12 Miscellaneous

10. For $i = 1$ and $i = 2$, the line L_i passes through the point $P_i(x_i, y_i, z_i)$ and is parallel to the vector $\mathbf{v}_i = \langle a_i, b_i, c_i \rangle$. Let L be the line through P_1 parallel to $\mathbf{v}_2$. Now, by definition, the lines L_1 and L_2 are skew if and only they are not coplanar. But L_1 and L_2 are coplanar if and only if the plane determined by L_1 and L is the same as the plane determined by L_2 and L; that is, if and only if the plane determined by L_1 and L is the same as the plane that contains the segment $P_1 P_2$ and the line L. Thus L_1 and L_2 are coplanar if and only if $\overrightarrow{P_1 P_2}$, $\mathbf{v}_1$, and $\mathbf{v}_2$ are coplanar, and this condition is equivalent (by a theorem from early in Chapter 13) to the condition

$$| \overrightarrow{P_1 P_2} \cdot \mathbf{v}_1 \times \mathbf{v}_2 | = 0.$$

That is, L_1 and L_2 are coplanar if and only if $\begin{vmatrix} x_1 - x_2 & y_1 - y_2 & z_1 - z_2 \\ a_1 & b_1 & c_1 \\ a_2 & b_2 & c_2 \end{vmatrix} = 0.$ This establishes

the conclusion given in Problem 10.

16. $x + 2y + 3z = 6$

22. Trajectory: $x = (v_0 \cos \alpha) t$, $y = -\frac{1}{2}gt^2 + (v_0 \sin \alpha) t$. Determine when $y = \frac{1}{3}x\sqrt{3}$; you should get impact time

$$t_1 = \frac{2v_0}{g} \left(\sin \alpha - \frac{\sqrt{3}}{3} \cos \alpha \right).$$

Now find the point of impact, and thus the range R as a function of the angle α: You should obtain

$$R = R(\alpha) = \tfrac{2}{3}\sqrt{3} \, (v_0 \cos \alpha) \, t_1.$$

Finally show that $dR/d\alpha = 0$ when $\tan 2\alpha = -\sqrt{3}$.

28. $\dfrac{dx}{dt} = \dfrac{dr}{dt} \cos \theta - (r \sin \theta) \dfrac{d\theta}{dt}$ and $\dfrac{dy}{dt} = \dfrac{dr}{dt} \sin \theta + (r \cos \theta) \dfrac{d\theta}{dt}$. Now show that

$$\left(\frac{dx}{dt} \right)^2 + \left(\frac{dy}{dt} \right)^2 + \left(\frac{dz}{dt} \right)^2 = \left(\frac{dr}{dt} \right)^2 + \left(r \left(\frac{d\theta}{dt} \right) \right)^2 + \left(\frac{dz}{dt} \right)^2.$$

34. Replace y by r to obtain $(r-1)^2 + z^2 = 1$. Simplify this to $r^2 - 2r + z^2 = 0$, then convert to the rectangular equation $x^2 + y^2 + z^2 = 2\sqrt{x^2 + y^2}$.

40. Because $z = ax^2 + by^2$ is a paraboloid (not a hyperboloid), we know that $ab > 0$. The projection into the xy-plane of the intersection K has the equation $ax^2 - Ax + by^2 - By = 0$. We may also assume that $a > 0$ and $b > 0$ (multiply through by -1 if necessary). It follows by completing squares that this projection is either empty, a single point, or an ellipse. In the latter case, it follows from Problem 32 that K is an ellipse.

46. The curvature is

$$\kappa(t) = \frac{|x'(t)y''(t) - x''(t)y'(t)|}{\left[(x'(t))^2 + (y(t))^2 \right]^{3/2}} = \frac{ab \sin^2 t + ab \cos^2 t}{(a^2 \sin^2 t + b^2 \cos^2 t)^{3/2}} = \frac{ab}{(a^2 \sin^2 t + b^2 \cos^2 t)^{3/2}}.$$

$$\kappa'(t) = \tfrac{3}{2}ab \left(a^2 \sin^2 t + b^2 \cos^2 t \right)^{-5/2} \left(2a^2 \sin t \, \cos t - 2b^2 \sin t \, \cos t \right) = \frac{3ab \, (\sin t \, \cos t) \, (a^2 - b^2)}{(a^2 \sin^2 t + b^2 \cos^2 t)^{5/2}}.$$

$\kappa'(t) = 0$ when $t = 0$, $\pi/2$, π, $3\pi/2$: $\kappa(0) = \dfrac{ab}{b^3} = \dfrac{a}{b^2} = \kappa(\pi)$, and $\kappa(\pi/2) = \dfrac{ab}{a^3} = \dfrac{b}{a^2} = \kappa(3\pi/2)$. Because $a > b$, κ is maximal when $t = 0$ and when $t = \pi$—at $(\pm a, 0)$— and minimal when $t = \pi/2$ and when $t = 3\pi/2$—at $(0, \pm b)$.

49. Because $y(x) = Ax + Bx^3 + Cx^5$ is an odd function, the condition $y(1) = 1$ will imply that $y(-1) = -1$. Because $y'(x) = A + 3Bx^2 + 5Cx^4$ is an even function, the condition $y'(1) = 0$ will imply that $y'(-1) = 0$ as well. Because the graph of y is symmetric about the origin, the condition that the curvature is zero at $(1,1)$ will imply that it is also zero at $(-1,-1)$. The curvature at (x,y) is

$$\frac{|6Bx + 20Cx^3|}{\left(1 + (A + 3Bx^2 + 5Cx^4)^2\right)^{3/2}}$$

by Problem 50 in Section 13.5, so the curvature at $(1,1)$ will be zero when $|6B + 20C| = 0$. Thus we obtain the simultaneous equations

$$
\begin{array}{rcrcrcl}
A &+& B &+& C &=& 1, \\
A &+& 3B &+& 5C &=& 0, \\
 && 3B &+& 10C &=& 0.
\end{array}
$$

These are easy to solve for $A = \frac{15}{8}$, $B = -\frac{5}{4}$, and $C = \frac{3}{8}$. Thus the equation of the connecting curve is $y = \frac{15}{8}x - \frac{5}{4}x^3 + \frac{3}{8}x^5$.

Chapter 13: Partial Differentiation
Section 13.2

1. The xy-plane

4. All points of the xy-plane other than the line $y = x$

7. All points (x, y) such that $x^2 + y^2 \leq 1$; that is, all points on and within the unit circle centered at the origin

10. The points in the plane for which $x^2 - y^2 > 1$. This set consists of the points strictly to the right of the right branch of the hyperbola $x^2 - y^2 = 1$ together with the points strictly to the left of the left branch of that hyperbola.

16. All points in space that lie strictly above (one might say "within") the paraboloid with equation $z = x^2 + y^2 = r^2$; this is a circular paraboloid with "vertex" at the origin, symmetric about the z-axis, and opening upward.

19. All points (x, y, z) of space for which $x^2 + y^2 < z$; that is, all points below the paraboloid with equation $x = x^2 + y^2$

22. The plane $z = x$, which passes through the y-axis at a $45°$ angle to the xy-plane.

28. The cylinder $z = 16 - y^2$ with rulings parallel to the x-axis and having the parabola $z = 16 - y^2$ as its trace in the yz-plane.

34. Some typical level curves of
$$f(x, y) = y - x^2$$
are shown at the right.

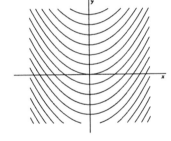

40. The level curves are circles centered at the origin.

46. The level surfaces are concentric circular cylinders with axis the y-axis and with rulings straight lines parallel to the y-axis.

49. Matches Fig. 13.2.30

52. Matches Fig. 13.2.29

55. Matches Fig. 13.2.42

58. Matches Fig. 13.2.43

Section 13.3

4. 0

10. 0

16. $\arcsin\left(-\frac{1}{2}\sqrt{2}\right) = \frac{1}{4}\pi$

22. The limit does not exist because the denominator is approaching zero and the numerator is not.

25. The limit is $\ln(1 + 0 + 0) = 0$.

28. Because $\ln x$ is continuous at $x = 1$ and $\sin x$ is continuous at $x = 0$, this limit has the value $\sin(\ln(1 + 0 + 0)) = 0$.

31. All points above (not "on") the graph of $y = -x$.

34. All points below (not "on") the graph of $y = 2x$.

40. This limit is 1 because $(\sin z)/z \to 1$ as $z \to 0$.

43. $\dfrac{x^2 - m^2x^2}{x^2 + m^2x^2}$ has limit $\dfrac{1 - m^2}{1 + m^2}$ as $x \to 0$, and hence the given limit does not exist.

46. If $y = z = 0$, then $\dfrac{x^2 + y^2 - z^2}{x^2 + y^2 + z^2} = 1$, but if $x = y = 0$, then $\dfrac{x^2 + y^2 - z^2}{x^2 + y^2 + z^2} = -1$. Hence the given limit does not exist.

49. $\dfrac{xy}{2x^2 + 3y^2} = \dfrac{m}{2 + 3m^2}$ on the line $y = mx$.

Section 13.4

4. $f_x(x, y) = (2x + x^2y)\, e^{xy}, \quad f_y(x, y) = x^3 e^{xy}$

10. $f_x(x, y) = \dfrac{y}{1 + x^2y^2}, \quad f_y(x, y) = \dfrac{x}{1 + x^2y^2}$

16. $\dfrac{\partial f}{\partial u} = (4u - 4u^3 - 6uv^2)\, e^{-u^2 - v^2}, \quad \dfrac{\partial f}{\partial v} = (6v - 4u^2v - 6v^3)\, e^{-u^2 - v^2}$

22. $z_{xy} = z_{yx} = 10x + 4y^3$

28. $z_{xy} = z_{yx} = (x^2y + xy^2)\sec xy + 2(x + y)\sec xy \tan xy + 2(x^2y + xy^2)\sec xy \tan^2 xy$

31. A normal to the plane at (x, y, z) is $\langle 2x, 2y, -1 \rangle$. So a normal to the plane at $(3, 4, 25)$ is $\langle 6, 8, -1 \rangle$. Therefore an equation of the tangent plane is $6x + 8y - z = 25$.

34. $2x + 2y - \pi z = 4 - \pi$

40. $3x - 4y - 5z = 0$

43. No such function f because $f_{xy} \neq f_{yx}$

46. Matches Fig. 13.4.17

49. Matches Fig. 13.4.15

52. Because $z_x = e^{x+y} = z_y$, all subsequent partial derivatives must also be equal to e^{x+y}.

55. You should find that $u_{xx} = -n^2 e^{-n^2 kt} \sin nx$.

58. (a) You should obtain $u_{xx} = \dfrac{y^2 - x^2}{(x^2 + y^2)^2}$.

   (b) $u_{xx} + u_{yy} = \dfrac{1}{u} \neq 0$, so this function does not satisfy Laplace's equation.

   (c) You should obtain $u_{xx} = \dfrac{2xy}{(x^2 + y^2)^2}$.

   (d) You should obtain $u_{xx} = e^{-x} \sin y$.

64. Let $P(a, b, c)$ be a point on the upper nappe of the cone, for which $z = z(x, y) = \sqrt{x^2 + y^2}$. Thus $c = \sqrt{a^2 + b^2}$. We assume that a and b are not both zero. It turns out that $z_x(P) = \dfrac{a}{c}$ and $z_y(P) = \dfrac{b}{c}$. So an equation of the tangent plane is

$$\frac{a(x - a)}{\sqrt{a^2 + b^2}} + \frac{b(y - b)}{\sqrt{a^2 + b^2}} - \left(z - \sqrt{a^2 + b^2}\right) = 0;$$

this can be simplified to $z = \dfrac{ax + by}{\sqrt{a^2 + b^2}}$. When $x = 0$ and $y = 0$, we have $z = 0$. Therefore the tangent plane at $P(a, b, c)$ passes through the origin. The same result for the lower nappe of the cone follows by symmetry.

Section 13.5

4. $(-1, 0, -1)$

10. $(1, -1, -1)$.

13. The only possible extremum is $f(1, 1) = 1$. Put $x = 1 + h$, $y = 1 + k$. Then $f(x, y) = h^2 + k^2 + 1$, so that $(1, 1)$ yields 1 as the global minimum value of f.

16. The surface clearly opens downward. There are two equally high global extrema, at $(1, 1, 2)$ and $(-1, -1, 2)$; there is a saddle point at $(0, 0, 0)$.

22. The only critical point occurs at $(0, 0)$. By an argument like the one in the preceding solution, $f(x, y)$ has the global maximum value 1 there.

28. Global maximum: 2, which occurs at the two points $(1, \sqrt{2})$ and $(1, -\sqrt{2})$. Global minimum: -2, which occurs at $(-1, \sqrt{2})$ and at $(-1, -\sqrt{2})$.

31. The square of the distance is
$$f(x, y) = (x - 7)^2 + (y + 7)^2 + (49 - 2x - 3y)^2;$$
$f(x, y)$ has only one critical point, at $(15, 5)$, and the point on the plane closest to $Q(7, -7, 0)$ is $(15, 5, 4)$.

34. The square of the distance is
$$f(x, y) = x^2 + y^2 + \frac{8}{x^4 y^8};$$
the only critical point of $f(x, y)$ in the first quadrant is $(1, \sqrt{2})$, and the point in the first octant and on the surface that is closest to $Q(0, 0, 0)$ is $(1, \sqrt{1}, 1/\sqrt{2})$.

37. If the dimensions of the box are x by y by z, then we are to minimize total surface area
$$A(x, y) = 2xy + 2(x + y)\frac{1000}{xy}$$
where $x > 0$ and $y > 0$. The only critical point of $A(x, y)$ is at $(10, 10)$, and by an argument like the one in Example 7, this yields the global minimum value of $A(x, y)$. When $x = 10 = y$, we find that $z = 10$, so the box of minimal total surface area is a cube measuring 10 in. along each edge.

40. Suppose that the base of the box measures x by y feet and that its height is z feet. Its top and bottom cost $\$3xy$ each, its front and back cost $\$4xz$ each, and its other two sides cost $\$yz$ each. Hence we are to minimize
$$f(x, y) = 6xy + 2 \cdot 4 \cdot (x + y)\frac{48}{xy}$$
where $x > 0$ and $y > 0$. The only critical point of $f(x, y)$ is at $(4, 4)$, and when $x = 4 = y$, we find that $z = 3$. Hence the least expensive such box has base 4 by 4 ft and its height is 3 ft.

43. Frontage: x. Depth: y. Height: z. We are to minimize cost $C = 2(2xz + xy) + 8yz$ given $xyz = 8000$. Begin by expressing C as a function of x and y alone.

46. We maximize volume $V(x, y) = cxy\left(1 - \frac{x}{a} - \frac{y}{b}\right)$; the critical points are $(0, 0)$, $(0, b)$, $(a, 0)$, and $\left(\frac{a}{3}, \frac{b}{3}\right)$. The maximum volume is $\frac{abc}{27}$, for $x = \frac{a}{3}$, $y = \frac{b}{3}$, and $z = \frac{c}{3}$.

49. Maximize $V = V(x, y) = 4xy\left(1 - x^2 - y^2\right)$ with domain the circular disk $x^2 + y^2 \leq 1$.

52. See the diagram at the right. We maximize the area
$$A(x, y) = xy + \frac{x}{4}\sqrt{(24 - x - 2y)^2 - x^2}.$$
Both partial derivatives are zero at points
$$(0, -12), \ (0, 4), \ \left(24\left(2 \pm \sqrt{3}\right), 4\left(3 \pm \sqrt{3}\right)\right).$$

The only possible point at which A is maximized is $\left(24\left(2 - \sqrt{3}\right), 4\left(3 - \sqrt{3}\right)\right)$; at that point the area is approximately 38.58 square feet.

55. See the diagram at the right. We minimize the area

$$A = xh + 2wy$$

given fixed volume

$$V = \tfrac{1}{2}xhy.$$

Using the Pythagorean theorem, we see that

$$h = \tfrac{1}{2}\sqrt{4w^2 - x^2}.$$

So we minimize

$$A\left(x, w\right) = \tfrac{1}{2}x\sqrt{4w^2 - x^2} + \frac{8Vw}{x\sqrt{4w^2 - x^2}} = \frac{x^2\left(4w^2 - x^2\right) + 16Vw}{2x\sqrt{4w^2 - x^2}}.$$

Both partial derivatives are zero when $x = 2^{5/6}V^{1/3}$, $w = 2^{1/3}V^{1/3}$, $y = 2^{1/3}V^{1/3}$, and $h = 2^{-1/6}V^{1/3}$. The minimum area is $2^{2/3} \cdot 3V^{2/3}$.

58. Suppose that the cubes have edges x, y, and z, respectively. We are to maximize and minimize $A = 6x^2 + 6y^2 + 6z^2$ given $x^3 + y^3 + z^3 = V$ (a constant). Note that x, y, and z are also nonnegative. Write

$$A = A\left(x, y\right) = 6x^2 + 6y^2 + 6\left(V - x^3 - y^3\right)^{2/3}.$$

Then $A_x\left(x, y\right) = 0 = A_y\left(x, y\right)$ at the following points:

$$x = 0, \quad y = 0, \quad z = V^{1/3} : \qquad A = 6V^{2/3}.$$
$$x = 0, \quad y = \left(V/2\right)^{1/3} = z : \qquad A = 12\left(V/2\right)^{2/3}.$$
$$y = 0, \quad x = \left(V/2\right)^{1/3} = z : \qquad A = 12\left(V/2\right)^{2/3}.$$
$$x = y = z = \left(V/3\right)^{1/3} : \qquad A = 18\left(V/3\right)^{2/3}.$$

The coefficients of $V^{2/3}$ in the above expressions for A are, in the same order, 6, 7.56, 7.56, and 8.65 (the last three are approximations). So, to obtain the maximum surface area, make three equal cubes, each of edge length $\left(V/3\right)^{1/3}$. A single cube of edge length $V^{1/3}$ has minimal surface area.

Section 13.6

4. $dw = \left(ye^{x+y} + xye^{x+y}\right) dx + \left(xe^{x+y} + xye^{x+y}\right) dy.$

10. $dw = ye^{uv} dx + xe^{uv} dy + xyve^{uv} + xyue^{uv} dv.$

16. $dw = \left(1 - 2p^2\right) qre^{-\left(p^2+q^2+r^2\right)} dp + \left(1 - 2q^2\right) pre^{-\left(p^2+q^2+r^2\right)} dq + \left(1 - 2r^2\right) pqe^{-\left(p^2+q^2+r^2\right)} dr.$

19. $df = -\dfrac{dx + dy}{\left(1 + x + y\right)^2}$, so $\Delta f \approx -\dfrac{0.02 + 0.05}{\left(1 + 3 + 6\right)^2} = -0.0007$ (the true value is approximately -0.000695).

22. $df = \dfrac{\left(x + y + z\right)yz - xyz}{\left(x + y + z\right)^2} dx + \dfrac{\left(x + y + z\right)xz - xyz}{\left(x + y + z\right)^2} dy + \dfrac{\left(x + y + z\right)xy - xyz}{\left(x + y + z\right)^2} dz.$ Take $x = 2$, $y =$
     3, $z = 5$, $dx = -0.02$, $dy = 0.03$, and $dz = -0.03$. Then $\Delta f \approx \tfrac{1}{10}\left(-0.24 + 0.21 - 0.09\right) = -0.012.$
     (The true value is approximately -0.012323.)

28. Let $w = f(x, y) = \dfrac{x^{1/3}}{y^{1/5}}$; $x = 27$, $y = 32$, $\Delta x = -2 = \Delta y$. Answer: $\dfrac{6401}{4320} \approx 1.4817$.

31. Think of y as implicitly defined as a function of x by means of the equation $2x^3 + 2y^3 = 9xy$. It follows that

$$\frac{dy}{dx} = \frac{3y - 2x^2}{2y^2 - 3x}$$

and therefore that the slope of the line tangent to the graph of $2x^3 + 2y^3 = 9xy$ at the point $(1, 2)$ is $\frac{4}{5}$. The straight line through $(1, 2)$ with that slope has equation

$$y = 2 + \tfrac{4}{5}(x - 1);$$

when $x = 1.1$, we have $y_{\text{curve}} - y_{\text{line}} = 2 + \frac{4}{5}(0.1) = 2.08$. The true value of y is approximately 2.0757625.

34. (a) The volume of the cylinder is $V(r, h) = \pi r^2 h$, so

$$dV(r, h, dr, dh) = 2\pi r h\, dr + \pi r^2\, dh.$$

With $r = 3$, $h = 9$, $dr = \pm 0.1$, and $dh = \pm 0.1$, we find that

$$dV = (117\pi) \cdot (\pm 0.1) = \pm(11.7)\pi \approx \pm 36.7566.$$

(b) The total surface area of the cylinder is $A(r, h) = 2\pi r h + 2\pi r^2$, so

$$dA(r, h, dr, dh) = (2\pi h + 4\pi r)\, dr + 2\pi r\, dh.$$

With $r = 3$, $h = 9$, $dr = \pm 0.1$, and $dh = \pm 0.1$, we find that

$$dA = \pm(3.6)\pi \approx \pm 11.3097.$$

So the maximum possible error in measuring the volume of the cylinder is approximately 36.7566 cm^3 and the maximum possible error in measuring its total surface area is approximately 11.3097 cm^2.

43. In part (b), you should find that $f_x(0, 0) = f_y(0, 0) = 0$.

Section 13.7

1. $\dfrac{dw}{dt} = \dfrac{dw}{dx} \cdot \dfrac{dx}{dt} + \dfrac{dw}{dy} \cdot \dfrac{dy}{dt} = -2xe^{-x^2 - y^2} - yt^{-1/2}e^{-x^2 - y^2} = -(2t + 1)e^{-t^2 - t}$.

4. $\dfrac{dw}{dt} = \dfrac{2t}{1 + t^2}$.

10. $r_x = y^2 + z^2 + 2(xy + yz + xz) - 4x - 2y - 2z;$
$r_y = x^2 + z^2 + 2(xy + yz + xz) - 2x - 4y - 2z;$
$r_z = x^2 + y^2 + 2(xy + yz + xz) - 2x - 2y - 4z.$

16. $\dfrac{\partial p}{\partial x} = \dfrac{\partial p}{\partial v}\dfrac{\partial v}{\partial x} + \dfrac{\partial p}{\partial w}\dfrac{\partial w}{\partial x}$,

$\dfrac{\partial p}{\partial y} = \dfrac{\partial p}{\partial v}\dfrac{\partial v}{\partial y} + \dfrac{\partial p}{\partial w}\dfrac{\partial w}{\partial y}$,

$\dfrac{\partial p}{\partial z} = \dfrac{\partial p}{\partial v}\dfrac{\partial v}{\partial z} + \dfrac{\partial p}{\partial w}\dfrac{\partial w}{\partial z}$,

$\dfrac{\partial p}{\partial t} = \dfrac{\partial p}{\partial v}\dfrac{\partial v}{\partial t} + \dfrac{\partial p}{\partial w}\dfrac{\partial w}{\partial t}$.

22. $z_x = -\dfrac{y^2}{5z^4 + y}$ and $z_y = -\dfrac{2xy + z}{5z^4 + y}$.

25. $w_x = 6x$ and $w_y = 6y$.

31. $x - y - z = 0$

34. Let $x = x(t)$ denote the length of each edge of the square base of the box and let $y = y(t)$ denote its height. Then its volume is $V = x^2 y$ and its surface area is $A = 2x^2 + 4xy$. Hence

$$\frac{dV}{dt} = 2xy\frac{dx}{dt} + x^2\frac{dy}{dt} \quad \text{and} \quad \frac{dA}{dt} = 4(x + y)\frac{dx}{dt} + 4x\frac{dy}{dt}.$$

Substitution of the given data $x = 100$ (cm), $y = 100$, $dx/dt = 2$, and $dy/dt = -3$ yields these results: At the time in question, $dV/dt = -10000$ (cm^3/min) and $dA/dt = 400$ (cm^2/min).

37. For this gas sample, we have $V = 10$ when $p = 2$ and $T = 300$. Substitution in the equation $pV = nRT$ yields $nR = \frac{1}{15}$. Moreover, with t in minutes, we have

$$V = \frac{nRT}{p}, \quad \text{so} \quad \frac{dV}{dt} = nR\left(\frac{1}{p}\frac{dT}{dt} - \frac{T}{p^2}\frac{dp}{dt}\right).$$

Now substitution of $nR = \frac{1}{15}$, $V = 10$, $p = 2$, $t = 200$, $dT/dt = 10$, and $dp/dt = 1$ yields the conclusion that the volume of the gas sample is decreasing at $\frac{14}{3}$ L/min at the time in question.

52. Show that $w_x = \dfrac{4xy^2}{(x^2 + y^2)^2} f'(u)$ and that $w_y = -\dfrac{4x^2 y}{(x^2 + y^2)^2} f'(u)$.

58. The result

$$T'(u, v, w) = \begin{bmatrix} a_1 & b_1 & c_1 \\ a_2 & b_2 & c_2 \\ a_3 & b_3 & c_3 \end{bmatrix}$$

follows immediately.

61. Here we compute

$$\begin{bmatrix} F_x & F_y & F_z \end{bmatrix} \begin{bmatrix} \sin\phi\cos\theta & \rho\cos\phi\cos\theta & -\rho\sin\phi\sin\theta \\ \sin\phi\sin\theta & \rho\cos\phi\sin\theta & \rho\sin\phi\cos\theta \\ \cos\phi & -\rho\sin\phi & 0 \end{bmatrix} = \begin{bmatrix} \dfrac{\partial q}{\partial\rho} & \dfrac{\partial q}{\partial\phi} & \dfrac{\partial q}{\partial\theta} \end{bmatrix},$$

which has first component

$$\frac{\partial q}{\partial\rho} = F_x\sin\phi\cos\theta + F_y\sin\phi\sin\theta + F_z\cos\phi,$$

second component

$$\frac{\partial q}{\partial\phi} = F_x\rho\cos\phi\cos\theta + F_y\rho\cos\phi\sin\theta - F_z\rho\sin\phi,$$

and third component

$$\frac{\partial q}{\partial\theta} = -F_x\rho\sin\phi\sin\theta + F_y\rho\sin\phi\cos\theta.$$

Section 13.8

1. $\langle 3, -7 \rangle$

4. $\langle \frac{1}{8}\pi\sqrt{2}, -\frac{3}{8}\pi\sqrt{2} \rangle$

10. $\langle 160, -240, 400 \rangle$

16. $\frac{4}{3}\sqrt{3}$

22. The maximum value is $\dfrac{1}{\sqrt{5}}$; it occurs in the direction $\langle 2, 1 \rangle$.

28. $\{\nabla f\}(7, 5, 5) = 19^{-1/2}\langle 2,\, 1,\, 2 \rangle$, a vector of magnitude $3/\sqrt{19}$. The latter is the maximal direction derivative of f, and the direction in which it occurs is $\langle 2, 1, 2 \rangle$. Because most people expect "direction" to be given by a *unit* vector, it is probably better to say that the direction is $\langle \frac{2}{3}, \frac{1}{3}, \frac{2}{3} \rangle$.

34. $4x + 23y + 17z = 25$

37. $\nabla\left(\dfrac{u}{v}\right) = \left\langle \dfrac{\partial}{\partial x}\left(\dfrac{u}{v}\right),\ \dfrac{\partial}{\partial y}\left(\dfrac{u}{v}\right) \right\rangle$

$= \left\langle \dfrac{u_x v - u v_x}{v^2},\ \dfrac{u_y v - u v_y}{v^2} \right\rangle = \dfrac{1}{v^2}\left(\langle u_x v,\, u_y v \rangle - \langle u v_x,\, u v_y \rangle \right)$

$= \dfrac{v\,\nabla u - u\,\nabla v}{v^2}.$

46. Let $F(x, y, z) = 2x^2 + 3y^2 - z$. The equation $F(x, y, z) = 0$ has the paraboloid as its graph. The vector $\mathbf{n} = \langle 4, -3, -1 \rangle$ is normal to the plane. $\nabla F = \langle 4x, 6y, -1 \rangle$, and we merely require that ∇F be parallel to $\mathbf{n}$. This leads to $x = 1$, $y = -\frac{1}{2}$, and (through $F(x, y, z) = 0$) to $z = \frac{11}{4}$. So an equation of the required plane is $4(x - 1) - 3\left(y + \frac{1}{2}\right) - \left(z - \frac{11}{4}\right) = 0$; that is, $16x - 12y - 4z = 11$.

52. (a) $\nabla z = \dfrac{\langle -0.06x, -0.14y \rangle}{\left(1 + (0.00003)x^2 + (0.00007)y^2\right)^2}$. The slope of this hill at the point $(100, 100, 500)$ in the

northwest direction is $\nabla z\,(100, 100) \cdot \dfrac{\sqrt{2}}{2}\langle -1, 1 \rangle = -\sqrt{2}$. Your initial rate of *descent* is $\sqrt{2}$ feet per foot, and the angle of descent is about $54°\,44'\,8''$.

(b) In the northeast direction, the initial rate of descent is $-\dfrac{5\sqrt{2}}{2}$, and the descent angle is about $74°\,12'\,25''$.

55. Let

$$f(x, y) = \tfrac{1}{1000}\left(3x^2 - 5xy + y^2\right).$$

Then

$$\nabla f(x, y) = \tfrac{1}{1000}\langle 6x - 5y,\, 2y - 5x \rangle,$$

and so $\nabla f(100, 100) = \langle \frac{1}{10}, -\frac{3}{10} \rangle$.

Part (a): A unit vector in the northeast direction is $\mathbf{u} = \langle 1, 1 \rangle/\sqrt{2}$, so the directional derivative of f at $(100, 100)$ in the northeast direction is

$$\langle \tfrac{1}{10}, -\tfrac{3}{10} \rangle \cdot \langle 1, 1 \rangle /\sqrt{2} = -\tfrac{1}{10}\sqrt{2}.$$

Hence you will initially be descending the hill, at an angle of $-\arctan\left(\frac{1}{10}\sqrt{2}\right)$, approximately $8.049467°$.

Part (b): A unit vector in the direction $30°$ north of east is $\mathbf{u} = \frac{1}{2}\langle \sqrt{3}, 1 \rangle$, so the directional derivative of f at $(100, 100)$ in the direction of $\mathbf{u}$ is

$$\langle \tfrac{1}{10}, -\tfrac{3}{10} \rangle \cdot \langle \sqrt{3}, 1 \rangle /2 = -\tfrac{1}{20}\left(3 - \sqrt{3}\right).$$

Hence you will initially be descending the hill, at an angle of approximately $3.627552°$.

58. $\mathbf{v}(t) = \left\langle \dfrac{dr}{dt}\cos\theta - \dfrac{d\theta}{dt}r\sin\theta, \ \dfrac{dr}{dt}\sin\theta + \dfrac{d\theta}{dt}r\cos\theta \right\rangle.$

Section 13.9

1.   The equations to be solved are

$$2 = 2\lambda x,$$

$$1 = 2\lambda y,$$

$$x^2 + y^2 = 1.$$

Maximum: $\sqrt{5}$, at $\left(\frac{2}{5}\sqrt{5}, \frac{1}{5}\sqrt{5}\right)$; minimum: $-\sqrt{5}$, at $\left(-\frac{2}{5}\sqrt{5}, -\frac{1}{5}\sqrt{5}\right)$

4. $\langle 2x, 2y \rangle = \lambda\langle 2,3\rangle$ leads to $x = \frac{12}{13}$, $y = \frac{18}{13}$. The minimum is $\frac{36}{13}$, and there is no maximum.

10. We obtain $\langle yz, xz, xy \rangle = \lambda\langle 2x, 2y, 2z\rangle$. There are fourteen solutions! If $xyz \neq 0$, take all possible combinations of sign in $\langle x, y, z \rangle$ where $x^2 = y^2 = z^2 = \frac{1}{3}$. If one of the variables is zero, the six solutions that result are $(1,0,0)$, $(-1,0,0)$, $(0,1,0)$, $(0,-1,0)$, $(0,0,1)$, and $(0,0,-1)$. Only the first eight lead to extrema. For the maximum of f, take either three plus signs or one to obtain $\frac{1}{9}\sqrt{3}$. To obtain the minimum, take two plus signs or none to obtain $-\frac{1}{9}\sqrt{3}$.

13. The equations to be solved are

$$2xy^2z^2 = 2\lambda x,$$

$$2x^2yz^2 = 8\lambda y,$$

$$2x^2y^2z = 18\lambda z,$$

$$x^2 + 4y^2 + 9z^2 = 27.$$

Solutions: $(0, y, z)$, $(x, 0, z)$, $(x, y, 0)$, and all eight of $(\pm 3, \pm\frac{3}{2}, \pm 1)$. Maximum: $81/4$, at all eight of the last critical points. Minimum 0, at all of the other critical points.

16. We let $g(x, y, z) = x^2 + y^2 - 1$ and $h(x, y, z) = 2x + 2y + z - 5$. The equation $\nabla f = \lambda\nabla g + \mu\nabla h$ then takes the form

$$\langle 0, 0, 1 \rangle = \lambda\langle 2x, 2y, 0\rangle + \mu\langle 2, 2, 1\rangle.$$

It follows that $\lambda x = -1 = \lambda y$; because $\lambda \neq 0$, $x = y$. So $x = y = \frac{1}{2}\sqrt{2}$ or $x = y = -\frac{1}{2}\sqrt{2}$. In the first case we see that $z = 5 - 2\sqrt{2}$; in the second, $z = 5 + 2\sqrt{2}$. It follows that the maximum is $5 + 2\sqrt{2}$ and that the minimum is $5 - 2\sqrt{2}$.

19. The equations to be solved are

$$2x = 3\lambda,$$

$$2y = 4\lambda,$$

$$3x + 4y = 100.$$

The only solution is $(12, 16)$. Clearly there can be no maximum, so $(12, 16)$ is the point on $3x + 4y = 100$ that is closest to the origin.

22. We are to minimize $(x-9)^2 + (y-9)^2 + (z-9)^2$ given $2x + 2y + z = 27$. The Lagrange multiplier equations are

$$2(x-9) = 2\lambda,$$
$$2(y-9) = 2\lambda,$$
$$2(z-9) = \lambda,$$
$$2x + 2y + z = 27.$$

The only solution is $(5,5,7)$, so this is the point on the given plane closest to $(9,9,9)$.

25. We are to minimize $x^2 + y^2 + z^2$ given $x^2 y^2 z = 4$. The Lagrange multiplier equations are

$$2x = 2\lambda xy^2 z,$$
$$2y = 2\lambda x^2 yz,$$
$$2z = \lambda x^2 y^2,$$
$$x^2 y^2 z = 4.$$

There are only four real solutions: $(\pm\sqrt{2}, \pm\sqrt{2}, 1)$. Although only one of these points lies in the first octant, all four yield the same global minimum value $\sqrt{5}$ for the distance from the given surface to $(0,0,0)$.

28. If the dimensions of the box are x by y by z, then we are to maximize xyz given the constraint $4x + 4y + 4z = 6$. The Lagrange multiplier equations are

$$yz = \lambda,$$
$$xz = \lambda,$$
$$xy = \lambda,$$
$$4x + 4y + 4z = 6.$$

There are four solutions: $\left(\frac{3}{2}, 0, 0,\right)$, $\left(0, \frac{3}{2}, 0\right)$, $\left(0, 0, \frac{3}{2}\right)$, and $\left(\frac{1}{2}, \frac{1}{2}, \frac{1}{2}\right)$. The last of these yields the maximum volume, $\frac{1}{8}$.

31. If the base of the box measures x by y and its height is z, then we are to minimize $6xy + 10xz + 10yz$ given $xyz = 600$. The Lagrange multiplier equations are

$$6y + 10z = \lambda yz,$$
$$6x + 10z = \lambda xz,$$
$$10x + 10y = \lambda xy,$$
$$xyz = 600.$$

The only real solution is $(10, 10, 6)$, so the least expensive such box has base 10 in. by 10 in. and height 6 in.

34. If the base of the box measures x by y and its height is z, then we are to minimize $3xy + 2xz + 2yz$ given the constraint $xyz = 12$. The Lagrange multiplier equations are

$$3y + 2z = \lambda yz,$$
$$3x + 2z = \lambda xz,$$
$$2x \pm 2y = \lambda xy,$$
$$xyz = 12.$$

The only real solution is $(2, 2, 3)$, so the least expensive such box has base 2 m by 2 m and height 3 m.

40. Maximize and minimize $f(x, y) = (x - 1)^2 + (y - 1)^2$ given $g(x, y) = 4x^2 + 9y^2 - 36 = 0$. The method yields

$$\langle 2(x - 1), 2(y - 1) \rangle = \lambda \langle 8x, 18y \rangle.$$

The (eventual) answer is that the nearest point—with all the following data approximate—is

$$(1.249987537, 1.818122492), \quad \text{at distance} \quad 0.8554637225;$$

the farthest point is

$$(-2.907018204, -0.494073841), \quad \text{at distance} \quad 4.182947273.$$

Comment: To obtain these answers, we had to eliminate the four points $(-3, 0)$, $(3, 0)$, $(0, 2)$, and $(0, -2)$. Then we solved the resulting equation $25x^4 - 90x^3 - 108x^2 + 810x - 729 = 0$ by numerical methods, and verified that it has only two real solutions.

46. Lagrange multiplier method: We maximize $(x - u)^2 + (y - v)^2 + (z - w)^2$ subject to the constraints

$x^2 + y^2 - z = 0$,

$u^2 + v^2 - w = 0$,

$x + y + z - 12 = 0$, and

$u + v + x - 12 = 0$.

Solve the simultaneous equations $\quad 2(x - u) = 2\lambda x + \mu$,
$$2(y - v) = 2\lambda y + \mu,$$
$$2(z - w) = -\lambda + \mu,$$
$$-2(x - u) = 2\eta u + \zeta,$$
$$-2(y - v) = 2\eta v + \zeta, \text{ and}$$
$$-2(z - w) = -\eta u + \zeta \quad \text{to obtain the four cases below.}$$

Case (1): If $\lambda = 1$: $u = v$ and $x = y$.

Case (2): If $\eta = 1$: $u = v$ and $x = y$.

Case (3): If $u = v$: $\lambda = 1$ or $x = y$, so $x = y$.

Case (4): If $x = y$: $u = v$.

Therefore $x = y$ and $u = v$. So since $x^2 + y^2 + x + y = 12$,

$$x^2 + x - 6 = 0;$$
$$x = -3 \quad \text{or} \quad x = 2.$$

Similarly, $u = -3$ or $u = 2$. Thus we obtain the two points $(2, 2, 8)$ and $(-3, -3, 18)$. So the center of the ellipse is $\left(-\frac{1}{2}, -\frac{1}{2}, 13\right)$. The other axis is perpendicular to this one and lies in the horizontal line through the center of the ellipse. A vector $\mathbf{V}$ parallel to that line satisfies $\langle -5, -5, 10 \rangle \cdot \mathbf{V} = 0$ and $\mathbf{V} = \langle a, b, 0 \rangle$. So $b = -a$. Choose $\mathbf{V} = \langle 1, -1, 0 \rangle$. This second line has symmetric equations

$$x - \tfrac{1}{2} = \tfrac{1}{2} - y, \quad z = 13.$$

So $x + y = 1$, $z = 13$. This line meets the paraboloid when $x^2 + (1 - x)^2 = 13$: at the points $(3, -2, 13)$ and $(-2, 3, 13)$. The first axis has length $\sqrt{150} = 5\sqrt{6} \approx 12.247$. The second has length $\sqrt{50} = 5\sqrt{2}$, so it is the minor axis. Answers: $\frac{5}{2}\sqrt{6}$ and $\frac{5}{2}\sqrt{2}$.

Note: A simpler solution, which does not use the method of Lagrange multipliers, is the following:

If θ is the angle between the xy-plane and the plane $x + y + z = 12$, then

$$\cos \theta = \frac{\langle 1,1,1 \rangle \cdot \langle 1,0,0, \rangle}{|\langle 1,1,1 \rangle| \, |\langle 1,0,0 \rangle|} = \frac{1}{\sqrt{3}},$$

so $\sec \theta = \sqrt{3}$.

The projection of the ellipse into the xy-plane has equation

$$x^2 + y^2 + x + y = 12;$$
$$(2x + 1)^2 + (2y + 1)^2 = 50;$$
$$\left(x + \tfrac{1}{2}\right)^2 + \left(y + \tfrac{1}{2}\right)^2 = \tfrac{25}{2}.$$

This is a circle with center $\left(-\tfrac{1}{2}, -\tfrac{1}{2}\right)$ and radius $\tfrac{5}{2}\sqrt{2}$. So the major semiaxis of the ellipse will have length $\left(\tfrac{5}{2}\sqrt{2}\right) \sec \theta = \tfrac{5}{2}\sqrt{6}$ and minor semiaxis of lenth $\tfrac{5}{2}\sqrt{2}$.

52. We are to find the global extrema of $(x-3)^2 + (y-2)^2$ given $4x^2 + 9y^2 = 36$. The Lagrange multiplier equations are

$$2(x - 3) = 8\lambda x,$$

$$2(y - 2) = 18\lambda y,$$

$$4x^2 + 9y^2 = 36.$$

There are only two real solutions, so the point of the ellipse closest to $(3, 2)$ is approximately $(2.35587, 1.23825)$, at approximate distance 2.66147, and the point farthest from $(3, 2)$ is approximately $(-2.88144, -0.55670)$, at approximate distance 2.93472.

55. We are to find the extrema of $x^2 + y^2 + z^2$ given

$$(x - 1)^2 + (y - 2)^2 + (z - 3)^2 = 36.$$

The Lagrange multiplier equations are

$$2x = 2\lambda(x - 1),$$

$$2y = 2\lambda(y - 2),$$

$$2z = 2\lambda(x - 3),$$

$$(x - 1)^2 + (y - 2)^2(z - 3)^2 = 36.$$

There are only two solutions. The point of the sphere that is closest to $(0, 0, 0)$ is approximately $(-0.60357, -1.20713, -1.81070)$, at approximate distance 2.25834; the point farthest from $(0, 0, 0)$ is approximately $(2.60357, 5.20713, 7.81070)$, at approximate distance 9.74166.

58. To simplify this problem, let's think of the way the Lagrange multiplier method works. We conclude that we need to find the points of the ellipsoid at which the normal vector is also normal to the given plane. The plane has normal $\langle 2, 3, 1 \rangle$, so such points (x, y, z) of the ellipsoid satisfy the equations

$$8x = 2\lambda,$$

$$18y = 3\lambda,$$

$$2z = \lambda,$$

$$4x^2 + 9y^2 + z^2 = 36.$$

The only solutions of these equations are

$$\left(-\sqrt{3},\, -\tfrac{2}{3}\sqrt{3},\, -2\sqrt{3}\right) \quad \text{and} \quad \left(\sqrt{3},\, \tfrac{2}{3}\sqrt{3},\, 2\sqrt{3}\right).$$

The first of these is farthest from the plane and the second is closest (as the nearest point is clearly in the first octant).

If we let (x, y, z) be a point on the ellipsoid and (u, v, w) be a point on the plane, and try to find the extrema of $(x - u)^2 + (y - v)^2 + (z - w)^2$ (as in the solution of Problem 57), the Lagrange multiplier equations are

$$2(x - u) = 8\lambda x,$$

$$-2(x - u) = 2\mu,$$

$$2(y - v) = 18\lambda y,$$

$$-2(y - v) = 3\mu,$$

$$2(z - w) = 2\lambda z,$$

$$-2(z - w) = \mu,$$

$$4x^2 + 9y^2 + z^2 = 36,$$

$$2u + 3v + w = 10.$$

Neither *Mathematica* 2.2 nor *Mathematica* 3.0 was able to solve this system, either exactly or approximately.

Section 13.10

1. $f(x, y) = 2(x + 1)^2 + (y - 2)^2 - 1$.

4. $y + 3 = 0 = x - 2$ at $(2, -3)$. $AC - B^2 = -1 < 0$, so f has no extrema.

7. $x^2 + y = 0 = x + y^2$ at $(0, 0)$ and at $(-1, -1)$. At the first critical point, $AC - B^2 = -9 < 0$; no extremum there. At $(-1, -1)$, $AC - B^2 = 27$ and $A = -6$. Therefore f has the local maximum value 4 at $(-1, -1)$.

10. No extremum at $(0, 0)$; local maximum at $(1, 1)$.

16. The equations $6x + 12y - 6 = 0 = 12x + 6y^2 + 6$ lead to $x = 1 - 2y$ and $y^2 - 4y + 3 = 0$, and thus $(-1, 1)$ and $(-5, 3)$ are the only two critical points. But $\Delta = 72y - 144$, so the first is not an extremum and the second yields a local minimum for f.

22. Local (indeed, global) minimum at $(0, 0)$; no extremum at either of the other two critical points $(0, 1)$ and $(0, -1)$.

28. Classification of the critical points:

(x, y)	A	B	C	Δ	Classification	z
$(0, 0)$	$2e$	0	$4e$	$8e^2$	Local minimum	0
$(0, 1)$	-2	0	-8	16	Local maximum	2
$(0, -1)$	-2	0	-8	16	Local maximum	2
$(1, 0)$	-4	0	2	-8	Saddle point	1
$(-1, 0)$	-4	0	2	-8	Saddle point	1

The following diagram gives a geometric summary of the critical points of the function.

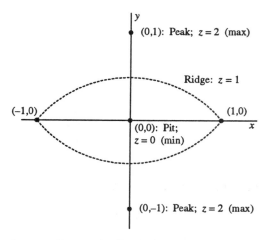

34. At the point with polar coordinates (r, θ), f takes on the value

$$\left(r^2 \sin \theta \cos \theta\right) \left(\cos^2 \theta - \sin^2 \theta\right) = \tfrac{1}{2} r^2 \sin 2\theta \cos 2\theta = \tfrac{1}{2} r^2 \sin 4\theta.$$

As you walk around the z-axis on this surface, the sine function gives you four high spots and four low spots, so the graph near the origin is another dog saddle.

Chapter 13 Miscellaneous

1. Transform the problem into polar form. Then the limit is $\lim\limits_{r \to 0} r^2 \sin \theta \cos \theta = 0$.

4. $g_x(0,0) = \lim\limits_{h \to 0} \dfrac{g(h,0) - g(0,0)}{h} = \lim\limits_{h \to 0} \dfrac{1}{h} \left(\dfrac{(h)(0)}{h^2 + 0^2} - 0 \right) = 0.$

7. The paraboloid is a level surface of $f(x, y, z) = x^2 + y^2 - z$ with gradient $\langle 2a, 2b, -1 \rangle$ at the point $\left(a, b, a^2 + b^2\right)$. The normal line through that point thus has parametric equations

$$x = 2at + a, \quad y = 2bt + b, \quad z = a^2 + b^2 - t.$$

Set $x = 0 = y$ and $z = 1$, solve the last equation for t, and substitute in the other two to find

$$2a \left(a^2 + b^2\right) - a = 0 = 2b \left(a^2 + b^2\right) - b.$$

It then follows that $a^2 + b^2 = \tfrac{1}{2}$ or $a = 0 = b$.

10. You should obtain $u_{xx} = \dfrac{4\pi^2}{(4\pi kt)^{5/2}} \left(x^2 - 2kt\right) e^{-x^2/4kt}.$

16. First find the equation of the tangent plane. Let (a, b, c) be the point of tangency; note that $abc = 1$. Let $F(x, y, z) = xyz - 1$; then $\nabla F = \langle yz, xz, xy \rangle$, so that a normal to the plane is

$$\nabla F(a, b, c) = \langle bc, ac, ab \rangle,$$

and the plane therefore has equation $bc(x - a) + ac(y - b) + ab(z - c) = 0$; that is, $bcx + acy + abz = 3$. Its intercepts are $x_0 = \dfrac{3}{bc}$, $y_0 = \dfrac{3}{ac}$, and $z_0 = \dfrac{3}{ab}$. By the formula $V = \tfrac{1}{3} hB$, its volume is $\tfrac{9}{2}$.

22. Write $w = w(u, v) = \displaystyle\int_u^v f(t)\, dt$ where $u = g(x)$, $v = h(x)$. Then $F(x) = w(u(x), v(x))$. Consequently $F'(x) = w_u u_x + w_v v_x = -f(u) g'(x) + f(v) h'(x) = f(h(x)) h'(x) - f(g(x)) g'(x)$.

25. $z = 500 - (0.003)\,x^2 - (0.004)\,y^2$. $\nabla z = \langle -(0.006)\,x, -(0.008)\,y \rangle$. $\nabla z\big|_{(-100,-100)} = \langle \frac{3}{5}, \frac{4}{5} \rangle$. So in order to maintain a constant altitude, you should move in the direction of either $\langle -4, 3 \rangle$ or $\langle 4, -3 \rangle$.

28. $(0, b)$, $(0, -b)$, $(a, 0)$, $(-a, 0)$.

34. See the diagram below.

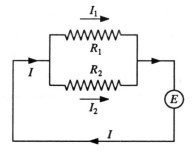

Let $x = I_1$ and $y = I_2$. Minimize $f(x, y) = R_1 x^2 + R_2 y^2$ given the constraint $g(x, y) = x + y - I = 0$ (I, R_1, and R_2 are all constants). The Lagrange multiplier method gives the vector equation $\langle 2R_1 x, 2R_2 y \rangle = \lambda \langle 1, 1 \rangle$. It follows without any difficulty that

$$I_1 = x = \frac{R_2 I}{R_1 + R_2} \quad \text{and} \quad I_2 = y = \frac{R_1 I}{R_1 + R_2}.$$

The resistance R of the parallel circuit satisfies the equation $E = IR$, but $E = R_1 I_1 = R_2 I_2$, so

$$IR = \frac{R_1 R_2 I}{R_1 + R_2};$$

$$R = \frac{R_1 R_2}{R_2 + R_2};$$

$$\frac{1}{R} = \frac{R_1 + R_2}{R_1 R_2} = \frac{1}{R_1} + \frac{1}{R_2}.$$

40. We seek the extrema of $f(x, y) = x^2 y^2$ given the constraint $g(x, y) = x^2 + 4y^2 - 24 = 0$. The Lagrange multiplier method gives the following results: The maximum value 36 of f occurs at the four points $(2\sqrt{3}, \sqrt{3})$, $(-2\sqrt{3}, \sqrt{3})$, $(2\sqrt{3}, -\sqrt{3})$, and $(-2\sqrt{3}, -\sqrt{3})$. The minimum value 0 of f occurs at the four points $(2\sqrt{6}, 0)$, $(-2\sqrt{6}, 0)$, $(0, \sqrt{6})$, and $(0, -\sqrt{6})$.

46. Global minimum -2 at $(1, -2)$ and at $(-1, -2)$; saddle point at $(0, -2)$.

Chapter 14: Multiple Integrals

Section 14.1

1. (a) $f(1,-2) \cdot 1 + f(2,-2) \cdot 1 + f(1,-1) \cdot 1 + f(2,-1) \cdot 1 + f(1,0) \cdot 1 + f(2,0) \cdot 1 = 198$

   (b) $f(2,-1) \cdot 1 + f(3,-1) \cdot 1 + f(2,0) \cdot 1 + f(3,0) \cdot 1 + f(2,1) \cdot 1 + f(3,1) \cdot 1 = 480$

4. $f\left(\frac{1}{2},\frac{1}{2}\right) + f\left(\frac{3}{2},\frac{1}{2}\right) + f\left(\frac{3}{2},\frac{1}{2}\right) + f\left(\frac{3}{2},\frac{3}{2}\right) = 4$

7. $\frac{1}{4}\pi^2 \left[f\left(\frac{1}{4}\pi,\frac{1}{4}\pi\right) + f\left(\frac{3}{4}\pi,\frac{1}{4}\pi\right) + f\left(\frac{1}{4}\pi,\frac{3}{4}\pi\right) + f\left(\frac{3}{4}\pi,\frac{3}{4}\pi\right) \right] = \frac{1}{2}\pi^2$

10. f is decreasing in the positive x-direction and the positive y-direction, and therefore $u \leqq M \leqq L$.

16. $\int_0^2 \left(\frac{56}{3}y^2 - 34\right) dy = -\frac{164}{9}$.

22. $\int_0^1 \frac{16}{3}e^y \, dy = \frac{16}{3}(e - 1)$.

28. $\int_1^e \frac{1}{x} \, dx = 1$.

31. $\int_{-1}^1 \int_{-2}^2 \left(2xy - 3y^2\right) dy \, dx = \int_{-1}^1 -16 \, dx = -32$. Other integral: $\int_{-2}^2 \int_{-1}^1 \left(2xy - 3y^2\right) dx \, dy$.

34. $\int_0^{\ln 2} \int_0^{\ln 3} e^x e^y \, dy \, dx = \int_0^{\ln 2} 2e^x \, dx = 2$.

37. If $0 \leq x \leq \pi$ and $0 \leq y \leq \pi$, then $0 \leqq f(x,y) \leqq \sin \pi = 1$. Hence every Riemann sum lies between $0 \cdot \text{area}(R)$ and $1 \cdot \text{area}(R)$.

Section 14.2

1. $\int_0^1 \left[y + xy\right]_0^x \, dx = \int_0^1 \left(x + x^2\right) dx = \frac{5}{6}$.

4. $\int_0^2 \left[\frac{1}{2}x^2 + xy\right]_{y/2}^1 \, dy = \int_0^2 \left(\frac{1}{2} + y - \frac{5}{8}y^2\right) dy = \frac{4}{3}$.

7. $\int_0^1 \int_x^{\sqrt{x}} (2x - y) \, dy \, dx = \int_0^1 \left[2xy - \frac{1}{2}y^2\right]_x^{\sqrt{x}} \, dx = \int_0^1 \left(2x^{3/2} - \frac{1}{2}x - 2x^2 + \frac{1}{2}x^2\right) dx$

   $= \left[\frac{4}{5}x^{5/2} - \frac{1}{4}x^2 - \frac{1}{2}x^3\right]_0^1 = \frac{1}{20}$.

10. 36

16. $\int_{-\sqrt{6}}^{\sqrt{6}} \int_{-4}^{2-x^2} x^2 \, dy \, dx = \int_{-\sqrt{6}}^{\sqrt{6}} 6x^2 - x^4 \, dx = \frac{48}{5}\sqrt{6}$.

19. $\int_0^\pi \int_0^{\sin x} x \, dy \, dx = \int_0^\pi x \sin x \, dx = \left[\sin x - x \cos x\right]_0^\pi = \pi$.

22. $\int_0^1 \int_0^{\sqrt{1-x^2}} xy \, dy \, dx = \int_0^1 \frac{1}{2}x(1 - x^2) \, dx = \frac{1}{8}$.

28. 0

34. $\sqrt{2} - 1$.

37. $\int_{-1.18921}^{1.18921} \int_{x^2-1}^{1/(1+x^2)} x \, dy \, dx = 0$.

40. $\int_0^{2.316934} \int_{x^2-2x}^{\sin x} x \, dy \, dx \approx 3.39454$.

43. Because $f(x,y) = xy$ is symmetric around the y-axis, the value of the integral is zero.

Section 14.3

1. The area is $A = \int_0^1 \int_{y^2}^y 1 \, dx \, dy = \frac{1}{6}$.

4. $A = \int_1^3 \int_{2x+3}^{6x-x^2} 1 \, dy \, dx = \frac{4}{3}$.

7. $A = \int_{-2}^2 \int_{2x^2-3}^{x^2+1} 1 \, dy \, dx = \frac{32}{3}$.

10. $\pi - \frac{2}{3}$

13. The volume is $V = \int_0^1 \int_0^2 (y + e^x) \, dy \, dx = \int_0^1 \left[\frac{1}{2}y^2 + ye^x\right]_0^2 dx$

$$= \int_0^1 (2 + 2e^x) \, dx = \left[2x + 2e^x\right]_0^1 = 2 + 2e - 2 = 2e.$$

16. $V = \int_0^2 \int_0^{4-2y} (3x + 2y) \, dx \, dy = \int_0^2 \left[\frac{3}{2}x^2 + 2xy\right]_0^{4-2y} dy = \int_0^2 (24 - 16y + 2y^2) \, dy = \frac{64}{3}$.

22. $\frac{837}{70}$

28. $\int_0^2 \int_{y/2}^{(4-y)/2} (8 - 4x - 2y) \, dx \, dy = \int_0^2 (9 - 8y + 2y^2) \, dy = \frac{16}{3}$.

31. $\int_{-1}^1 \int_{-\sqrt{1-x^2}}^{\sqrt{1-x^2}} (x + 1) \, dy \, dx = \int_{-1}^1 2(x + 1)\sqrt{1 - x^2} \, dx = \pi$.

34. $\int_{-1}^1 \int_{-\sqrt{1-x^2}}^{\sqrt{1-x^2}} \left[\sqrt{2 - x^2 - y^2} - (x^2 + y^2)\right] dy \, dx \approx 2.258652488356396218$.

37. The volume is $V = \int_0^1 \int_0^{\sqrt{1-y^2}} \sqrt{1 - y^2} \, dx \, dy$.

40. $V = \frac{4}{3}\pi abc$

Section 14.4

1. $A = \int_0^{2\pi} \int_0^1 r \, dr \, d\theta$.

4. $A = \int_{-\pi/4}^{\pi/4} \int_0^{2\cos 2\theta} r \, dr \, d\theta = \pi/2$.

7. $A = 2 \int_{\pi/6}^{\pi/2} \int_0^{1-2\sin\theta} r \, dr \, d\theta$.

10. $\frac{3}{2}\pi$

13. $\int_0^{\pi/2} \int_0^1 \frac{r}{1 + r^2} \, dr \, d\theta$

16. $\int_{\pi/4}^{\pi/2} \int_0^{\csc\theta} r^3 \cos^2\theta \, dr \, d\theta = \frac{1}{12}$.

19. The volume is $V = \int_0^{2\pi} \int_0^1 (2 + r\cos\theta + r\sin\theta) \, r \, dr \, d\theta = \int_0^{2\pi} \left[r^2 + \frac{1}{3}r^3\cos\theta + \frac{1}{3}r^3\sin\theta\right]_0^1 d\theta$

$$= \int_0^{2\pi} (1 + \frac{1}{3}\cos\theta + \frac{1}{3}\sin\theta) \, d\theta = \left[\theta + \frac{1}{3}\sin\theta - \frac{1}{3}\cos\theta\right]_0^{2\pi} = 2\pi.$$

22. $V = \displaystyle\int_0^{2\pi} \int_0^{1+\cos\theta} (1 + r\cos\theta)\, r\, dr\, d\theta = \frac{1}{6}\int_0^{2\pi} (3 + 9\cos^2\theta + 2\cos^4\theta)\, d\theta = \frac{11}{4}\pi.$

28. Domain of integration: the disk $x^2 + y^2 \le 1$. $\quad V = \displaystyle\int_0^{2\pi} \int_0^1 r(1 - r^2)\, dr\, d\theta = \frac{1}{2}\pi.$

31. $V = \displaystyle\int_0^{\pi/2} \int_0^{\sqrt{2\sin 2\theta}} r^3\, dr\, d\theta.$

34. The polar form of the integral is $\displaystyle\int_0^{\pi/2} \int_0^\infty \frac{r}{(1+r^2)^2}\, dr\, d\theta.$

37. $\displaystyle\int_0^{2\pi} \int_0^2 (4 - r^2)\, r\, dr\, d\theta = 8\pi.$

40. $\displaystyle\int_0^{2\pi} \int_0^2 \left(2\sqrt{20 - r^2} - 2r^2 \right) r\, dr\, d\theta = \int_0^{2\pi} \left[-\tfrac{1}{4}r^2 - \tfrac{2}{3}(20 - r^2)^{3/2} \right]_0^2 d\theta$

$\qquad = \displaystyle\int_0^{2\pi} \frac{80\sqrt{5} - 152}{3}\, d\theta = \tfrac{2}{3}\pi \left(80\sqrt{5} - 152 \right) \approx 56.308730092.$

Section 14.5

4. $M_y = \displaystyle\int_0^3 x(3 - x)\, dx = \frac{9}{2}.$

The area of the region is the same, so $\bar{x} = 1$. By symmetry, $\bar{y} = 1$ as well. Answer: $(1, 1)$.

10. By symmetry, $\bar{x} = 0$. $M_x = \frac{1}{2}\displaystyle\int_{-2}^2 (x^2 + 1)^2\, dx = \frac{206}{15}.$ The area of the region is $\frac{28}{3}$, so $\bar{y} = \frac{103}{70}.$

16. $M = \displaystyle\int_0^1 \int_{x^2}^{\sqrt{x}} (x^2 + y^2)\, dy\, dx = \frac{6}{35}, \quad M_y = \int_0^1 \int_{x^2}^{\sqrt{x}} (x^3 + xy^2)\, dy\, dx = \frac{55}{504}.$

Hence $\bar{x} = \frac{275}{432}.$ By symmetry, $\bar{y} = \bar{x}.$

22. We assume that $a > 0$. Then

$$M = \int_0^a \int_0^{a-x} (x^2 + y^2)\, dy\, dx = \tfrac{1}{6}a^4, \quad M_x = \int_0^a \int_0^{a-x} (x^2 y + y^3)\, dy\, dx = \tfrac{1}{15}a^5.$$

Because of the symmetry of x, y, and density in the region, it follows that $M_y = M_x$. Therefore $\bar{x} = \frac{2}{5}a = \bar{y}.$

28. $M = \frac{5}{3}\pi, \ M_x = 0, \ M_y = \frac{7}{4}\pi, \quad \bar{x} = \frac{21}{20}, \ \bar{y} = 0.$

34. $\sqrt{3} + \frac{4}{3}\pi \approx 5.90841.$

40. $M = k\pi$ by inspection. With the aid of Formula 113 from the endpapers of the text, $I_y = \frac{5}{4}k\pi$ and $I_x = \frac{1}{4}k\pi.$ Therefore $\hat{x} = \frac{1}{2}\sqrt{5}$ and $\hat{y} = \frac{1}{2}.$

43. Because $y = \sqrt{r^2 - x^2}$, $1 + \left(\dfrac{dy}{dx}\right)^2 = \dfrac{r^2}{r^2 - x^2}.$ So $M_y = \displaystyle\int_0^r x\,\frac{r}{\sqrt{r^2 - x^2}}\, dx = r^2.$ The length of the quarter-circle is $\frac{1}{2}\pi r$, so $\bar{x} = 2r/\pi.$ By symmetry, $\bar{y} = \bar{x}.$

46. $V = \left(2\pi\dfrac{r}{3} \right) \left(\tfrac{1}{2}rh \right) = \tfrac{1}{3}\pi r^2 h.$

49. With the same notational change as in the solution to Problem 48, the radius of revolution is $\frac{1}{2}(r + s)$, so the lateral area is $A = 2\pi \left(\dfrac{r + s}{2} \right) \sqrt{(s - r)^2 + h^2} = \pi(r + s)\, L.$

52. (a) $A = \displaystyle\int_0^h \sqrt{2py}\, dy = \frac{2}{3}h^{3/2}\sqrt{2p}.$ Now $r^2 = 2ph$, so $A = \frac{2}{3}h\sqrt{2ph} = \frac{2}{3}rh.$

$$M_y = \int_0^h \tfrac{1}{2}(2py)\, dy = \tfrac{1}{2}ph^2, \text{ so } \bar{x} = \frac{ph^2/2}{2rh/3} = \frac{3ph}{4r}. \text{ But } ph = \tfrac{1}{2}r^2, \text{ so } \bar{x} = \tfrac{3}{8}r.$$

(b) $V = 2\pi \bar{x}A = 2\pi M_y = \pi ph^2$. But $ph = \tfrac{1}{2}r^2$, so $V = \tfrac{1}{2}\pi r^2 h$.

58. Mass:

$$M = \int_0^\pi \int_0^{2\sin\theta} r^3 \sin\theta\, dr\, d\theta = \int_0^\pi 4\sin^5\theta\, d\theta = \frac{64}{15};$$

$$M_y = \int_0^\pi \int_0^{2\sin t} r^4 \sin\theta \cos\theta\, dr\, d\theta = \int_0^\pi \tfrac{32}{5}\sin^6\theta \cos\theta\, d\theta = 0;$$

$$M_x = \int_0^\pi \int_0^{2\sin t} r^4 \sin^2\theta\, dr\, d\theta = \int_0^\pi \tfrac{32}{5}\sin^7\theta\, d\theta = \frac{1024}{175};$$

$$\bar{x} = 0, \ \bar{y} = \frac{48}{35}.$$

Section 14.6

1. $\displaystyle\iiint_T f(x,y,z)\, dV = \int_0^1 \int_0^3 \int_0^2 (x + y + z)\, dx\, dy\, dz = \int_0^1 \int_0^3 \left[\tfrac{1}{2}x^2 + xy + xz\right]_0^2 dy\, dz$

$$= \int_0^1 \int_0^3 (2 + 2x + 2y)\, dy\, dz = \int_0^1 \left[2y + y^2 + xyz\right]_0^3 dz$$

$$= \int_0^1 (15 + 6z)\, dz = \left[15z + 3z^2\right]_0^1 = 18.$$

4. $\displaystyle\int_{-2}^6 \int_0^2 (4 + 4y + 4z)\, dy\, dz = \int_{-2}^6 (8 + 8z)\, dz = \left[8z + 4z^2\right]_{-2}^6 = 192.$

10. $\displaystyle\int_{-1}^1 \int_{-2}^2 \int_{y^2}^{8-y^2} z\, dz\, dy\, dx = \int_{-1}^1 \int_{-2}^2 (32 - 8y^2)\, dy\, dx = \int_{-1}^1 (128 - \tfrac{128}{3})\, dx = \tfrac{512}{3}.$

16. $\frac{128}{3}$

22. $M_{yz} = 0 = M_{xz}; \ \ M_{xy} = \tfrac{1}{4}\pi R^4.$

28. $\tfrac{1}{2}\pi R^4 H$

34. $M = ka^5, \ M_{yz} = \tfrac{7}{12}ka^6, \ \bar{x} = \bar{y} = \bar{z} = \tfrac{7}{12}a.$

40. $\frac{16}{45}$

46. Mass:

$$M = \int_{-1}^2 \int_{y^2}^{y+2} \int_{-\sqrt{z-y^2}}^{\sqrt{z-y^2}} 1\, dx\, dz\, dy$$

$$= \int_{-1}^2 \int_{y^2}^{y+2} 2\sqrt{z - y^2}\, dz\, dy$$

$$= \int_{-1}^2 \tfrac{4}{3}(2 + y - y^2)^{3/2}\, dy = \frac{81\pi}{32}.$$

By symmetry, $\bar{x} = 0$. Next,

$$M_{xz} = \int_{-1}^{2} \int_{y^2}^{y+2} \int_{-\sqrt{z-y^2}}^{\sqrt{z-y^2}} y \, dx \, dz \, dy$$

$$= \int_{-1}^{2} \int_{y^2}^{y+2} 2y\sqrt{z-y^2} \, dz \, dy$$

$$= \int_{-1}^{2} \tfrac{4}{3}y(2+y-y^2)^{3/2} \, dy = \frac{81\pi}{64},$$

and hence $\bar{y} = \tfrac{1}{2}$. Finally,

$$M_{xy} = \int_{-1}^{2} \int_{y^2}^{y+2} \int_{-\sqrt{z-y^2}}^{\sqrt{z-y^2}} z \, dx \, dz \, dy$$

$$= \int_{-1}^{2} \int_{y^2}^{y+2} 2z\sqrt{z-y^2} \, dz \, dy$$

$$= \int_{-1}^{2} \tfrac{4}{15}(12+12y+y^2-y^3-2y^4)\sqrt{2+y-y^2} \, dy = \frac{567\pi}{128},$$

and therefore $\bar{z} = \tfrac{7}{4}$. Therefore the centroid of the parabolic segment of Example 5 is located at the point $\left(0, \tfrac{1}{2}, \tfrac{7}{4}\right)$.

49. Because the centroid of T is at the point $C\left(\tfrac{1}{2}, \tfrac{1}{2}, \tfrac{1}{2}\right)$, the average of the squared distance of points of the cube T from C is

$$\bar{d} = \frac{1}{1} \int_{0}^{1} \int_{0}^{1} \int_{0}^{1} \left[(x-\tfrac{1}{2})^2 + (y-\tfrac{1}{2})^2 + (z-\tfrac{1}{2}) \right] dz \, dy \, dx$$

$$= \int_{0}^{1} \int_{0}^{1} (x^2 + y^2 - x - y + \tfrac{7}{12}) \, dy \, dx$$

$$= \int_{0}^{1} (x^2 - x + \tfrac{5}{12}) \, dx = \frac{1}{4}.$$

52. Using the result in the text that the centroid of the pyramid T is located at $\left(\tfrac{2}{5}, \tfrac{3}{5}, \tfrac{12}{5}\right)$ and the result of Problem 47 that T has volume $V = 6$, we find that the average of the square distance of points of T from its centroid is

$$\bar{d} = \frac{1}{V} \int_{0}^{2} \int_{0}^{(6-3x)/2} \int_{0}^{6-3x-2y} \left[(x-\tfrac{2}{5})^2 + (y-\tfrac{3}{5})^2 + (z-\tfrac{12}{5})^2 \right] dz \, dy \, dx$$

$$= \frac{1}{6} \int_{0}^{2} \int_{0}^{(6-3x)/2} \left(\tfrac{582}{25} - \tfrac{1131}{25}x + \tfrac{204}{5}x^2 - 12x^3 - \tfrac{854}{25}y \right.$$

$$\left. + \tfrac{242}{5}xy - 20x^2y + \tfrac{114}{5}y^2 - 15xy^2 - \tfrac{14}{3}y^3 \right) dy \, dx$$

$$= \frac{1}{6} \int_{0}^{2} \left(\tfrac{1341}{50} - \tfrac{1323}{25}x + \tfrac{1467}{25}x^2 - \tfrac{162}{5}x^3 + \tfrac{207}{32}x^4 \right) dx = \frac{67}{25}.$$

Section 14.7

1. $V = \int_0^{2\pi} \int_0^2 \int_{r^2}^4 r \, dz \, dr \, d\theta = 2\pi \int_0^2 \left(4r - r^3\right) \, dr = 2\pi \left[2r^2 - \tfrac{1}{4}r^4\right]_0^2 = 8\pi.$

4. Use the sphere $r^2 + z^2 \le a^2$, $0 \le \theta \le 2\pi$. We find its moment of inertia about the z-axis:

$$I_z = \int_0^{2\pi} \int_0^a \int_{-\sqrt{a^2-r^2}}^{\sqrt{a^2-r^2}} r^3 \, dz \, dr \, d\theta = 2\pi \int_0^a 2r^3 \sqrt{a^2 - r^2} \, dr.$$

Let $r = a \sin u$. This substitution yields

$$I_z = 4\pi a^5 \int_0^{\pi/2} \left(\cos^2 u - \cos^4 u\right)(\sin u) \, du = 4\pi a^5 \left[\tfrac{1}{5}\cos^5 u - \tfrac{1}{3}\cos^3 u\right]_0^{\pi/2} = \tfrac{8}{15}\pi a^5.$$

The answer may also be written in the form $I_z = \tfrac{2}{5}Ma^5$ where M is the mass of the sphere.

7. Mass: $M = \int_0^{2\pi} \int_0^a \int_0^h rz \, dz \, dr \, d\theta = \pi \int_0^a rh^2 \, dr = \left[\tfrac{1}{2}\pi r^2 h^2\right]_0^a = \tfrac{1}{2}\pi a^2 h^2.$

10. The diagram at the right shows the intersection of the xy-plane with the sphere and the cylinder. The plane and sphere intersect in the curve $r = 2$ and the cylindrical equation of the sphere is $r^2 + z^2 = 4$. The plane intersects the cylinder in the curve $r = 2\cos\theta$. Thus the volume V of their intersection is given by

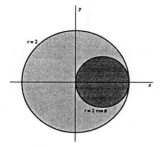

$$V = 4\int_0^{\pi/2} \int_0^{2\cos\theta} \int_0^{\sqrt{4-r^2}} r \, dz \, dr \, d\theta$$

$$= 4\int_0^{\pi/2} \int_0^{2\cos\theta} r\sqrt{4-r^2} \, dr \, d\theta$$

$$= 4\int_0^{\pi/2} \left[-\tfrac{1}{3}\left(4-r^2\right)^{3/2}\right]_0^{2\cos\theta} d\theta$$

$$= \tfrac{32}{3} \int_0^{\pi/2} \left(1 - \sin^3\theta\right) d\theta = \tfrac{16}{9}\left(3\pi - 4\right).$$

16. Let k denote the density of the cylinder and h its height. Then $m = \pi a^2 hk$. We choose coordinates so that the z-axis is the axis of symmetry of the cylinder and its base rests on the xy-plane. Then

$$I_z = \int_0^{2\pi} \int_0^a \int_0^h kr^3 \, dz \, dr \, d\theta = \tfrac{1}{2}\pi a^4 hk = \left(\pi a^2 hk\right)\left(\tfrac{1}{2}a^2\right) = \tfrac{1}{2}ma^2.$$

22. Locate the hemisphere with its base in the xy-plane and symmetric about the z-axis. It follows that $M = \tfrac{1}{4}k\pi a^4$ and $M_{xy} = \tfrac{2}{15}\pi ka^5$; its centroid is at $\left(0, 0, \tfrac{8}{15}a\right)$.

25. $M_{xy} = \tfrac{1}{8}\pi a^4$

28. $I_z = \tfrac{1016}{21}\pi a^7$

34. The moment of inertia of the ice-cream cone around the z-axis is

$$\Phi_z = \int_0^{2\pi} \int_0^{\pi/6} \int_0^{2a\cos\phi} (\rho\cos\phi)(\rho\sin\phi)^2 \rho^2 \sin\phi \, d\rho \, d\phi \, d\theta$$

$$= \int_0^{2\pi} \int_0^{\pi/6} \tfrac{32}{3} a^6 \cos^7 \phi \sin^3 \phi \, d\phi \, d\theta$$

$$= \int_0^{2\pi} \tfrac{47}{480} a^6 \, d\theta = \frac{47\pi a^6}{240}.$$

37. The value of the integral in part (a) is

$$\int_0^{2\pi} \int_0^{\pi} \int_0^{a} \exp\left(-\rho^3\right) \rho^2 \sin\phi \, d\rho \, d\phi \, d\theta = \int_0^{2\pi} \int_0^{\pi} \tfrac{1}{3} \left[1 - \exp(-a^3)\right] \sin\phi \, d\phi \, d\theta$$

$$= \int_0^{2\pi} \tfrac{2}{3} \left[1 - \exp(-a^3)\right] \, d\theta = \tfrac{4}{3}\pi \left(1 - e^{-a^3}\right).$$

Part (b): It is now clear that the limit of the last expression as $a \to \infty$ is $\tfrac{4}{3}\pi$.

40. We use the ball of radius a and volume $V = \tfrac{4}{3}\pi a^3$ centered at the point with Cartesian coordinates $(0,0,a)$ on the positive z-axis. The boundary of this ball is the spherical surface described by

$$\rho = 2a \cos\phi, \qquad 0 \leq phi \leq \tfrac{1}{2}\pi, \qquad 0 \leq \theta \leq 2\pi.$$

The average distance of points of this ball from the origin (its "south pole") is then

$$\overline{d} = \frac{1}{V} \int_0^{2\pi} \int_0^{\pi/2} \int_0^{2a \cos\phi} \rho^3 \sin\phi \, d\rho \, d\phi \, d\theta$$

$$= \frac{1}{V} \int_0^{2\pi} \int_0^{\pi/2} 4a^4 \cos^4 \phi \sin\phi \, d\phi \, d\theta$$

$$= \frac{1}{V} \int_0^{2\pi} \tfrac{4}{5} a^4 \, d\theta = \frac{8\pi a^4}{5V} = \frac{6a}{5}.$$

Section 14.8

1. $\sqrt{1 + (z_x)^2 + (z_y)^2} = \sqrt{11}$. The area of the ellipse is 6π, so the answer is $6\pi\sqrt{11}$.

4. $\sqrt{1 + (z_x)^2 + (z_y)^2} = \sqrt{1 + x^2}$. So the answer is

$$\int_0^1 \int_0^x \sqrt{1 + x^2} \, dy \, dx = \int_0^1 x\sqrt{1 + x^2} \, dx = \left[\tfrac{1}{3} \left(1 + x^2\right)^{3/2} \right]_0^1 = \tfrac{1}{3}\left(2\sqrt{2} - 1\right) \approx 0.609476.$$

10. $A = \tfrac{1}{6}\pi \left(17\sqrt{17} - 1\right) \approx 36.1769.$

16. $A = 2a^2 \left(\pi - 2\right)$, about 18% of the total surface area.

22. The area in question is given by

$$A = \int_0^{2\pi} \int_0^{\pi/6} a^2 \sin\phi \, d\phi \, d\theta = \left[2\pi - a^2 \cos\phi \right]_0^{\pi/6} = 2\pi a^2 \left(1 - \tfrac{1}{3}\sqrt{3}\right),$$

about 21.13% of the total surface area.

28. a) The surface area is

$$\int_{-1}^1 \int_{-1}^1 \frac{2}{\sqrt{4 - x^2 - y^2}} \, dy \, dx \approx 17.6411.$$

b) Taking advantage of symmetry, the surface area is

$$4 \int_0^1 \int_0^{1-x} \frac{2}{\sqrt{4 - x^2 - y^2}} \, dy \, dx \approx 2.09159.$$

Section 14.9

4. $x = \frac{1}{2}\sqrt{u+v}$, $y = \frac{1}{2}\sqrt{u-v}$, and $J_T = \dfrac{\partial(x,y)}{\partial(u,v)} = -\dfrac{1}{8\sqrt{u^2-v^2}}$.

7. $J_T = \dfrac{\partial(x,y)}{\partial(u,v)} = -\frac{1}{5}$.

10. First, $x = (u^2v)^{-1/3}$, $y = (uv^2)^{-1/3}$, and $J_T = \dfrac{\partial(x,y)}{\partial(u,v)} = \dfrac{1}{3u^2v^2}$. It follows that the area of the region is $\frac{1}{8}$.

13. The Jacobian is $6r$.

16. $r = \sqrt{z}$, $t = \sqrt{\dfrac{z}{x^2+y^2}}$, and $\theta = \arctan\left(\dfrac{y}{x}\right)$. The Jacobian of the transformation is $2\,(r/t)^3$. The volume of R turns out to be $45\pi/8$.

19. The spherical integral is $\displaystyle\lim_{a\to\infty}\int_0^{2\pi}\int_0^{\pi}\int_0^{a} \rho^3 e^{-k\rho^2}\sin\phi\,d\rho\,d\phi\,d\theta$.

22. The transformation $u = xy$, $v = y/x$ yields $x = \sqrt{u/v}$, $y = \sqrt{uv}$, and

$$\frac{\partial(u,v)}{\partial(x,y)} = \begin{vmatrix} y & x \\ -\dfrac{y}{x^2} & \dfrac{1}{x} \end{vmatrix} = \frac{2y}{x} = 2v.$$

It follows that the area of the region is

$$\int_1^2\int_1^2 2v\,dv\,du = 3,$$

$$M_x = \int_1^2\int_1^2 2u^{1/2}v^{3/2}\,d\,du = \tfrac{8}{15}\left(17 - 6\sqrt{2}\right),$$

$$M_y = \int_1^2\int_1^2 2u^{1/2}v^{1/2}\,dv\,du = \tfrac{8}{9}\left(9 - 4\sqrt{2}\right),$$

and hence $\overline{x} = \tfrac{8}{27}\left(9 - 4\sqrt{2}\right) \approx 0.990562$ and $\overline{y} = \tfrac{8}{45}\left(17 - 6\sqrt{2}\right) \approx 1.513728$.

25. The transformation $x = a\rho\sin\phi\cos\theta$, $y = b\rho\sin\phi\sin\theta$, $z = c\rho\cos\phi$ yields

$$\frac{\partial(x,y,z)}{\partial(\rho,\phi,\theta)} = \begin{vmatrix} a\sin\phi\cos\theta & a\rho\cos\phi\cos\theta & -a\rho\sin\phi\sin\theta \\ b\sin\phi\sin\theta & b\rho\cos\phi\sin\theta & b\rho\sin\phi\cos\theta \\ c\cos\phi & -c\rho\sin\phi & 0 \end{vmatrix} = abc\rho^2\sin\phi.$$

Now $x^2 + y^2 = (\rho^2\sin^2\phi)(a^2\cos^2\theta + b^2\sin^2\theta)$, and hence

$$I_z = \int_0^{2\pi}\int_0^{\pi}\int_0^{1} (\delta abc\rho^4\sin^3\phi)(a^2\cos^2\theta + b^2\sin^2\theta)\,d\rho\,d\phi\,d\theta = \tfrac{4}{15}\pi\delta abc(a^2 + b^2)$$

$$= \tfrac{1}{5}M(a^2 + b^2)$$

where M is the mass of the ellipsoid. By symmetry,

$$I_x = \tfrac{1}{5}M(b^2 + c^2) \quad\text{and}\quad I_y = \tfrac{1}{5}M(a^2 + c^2).$$

Chapter 14 Miscellaneous

1. $\displaystyle\int_0^1 \int_0^{x^3} (1+x^2)^{-1/2}\, dy\, dx$

4. $\displaystyle\int_0^4 \int_0^{y^{3/2}} x \cos y^4\, dx\, dy$

7. $V = \displaystyle\int_0^1 \int_u^{2-y} (x^2+y^2)\, dx\, dy.$

10. The domain of the integral will be the elliptical region bounded by the ellipse whose equation is $(x/2)^2 + y^2 = 1$. Let us use the transformation $x = 2r\cos\theta$, $y = r\sin\theta$. The Jacobian of this transformation is $2r$. The difference in the z-values of the two paraboloids turns out to be $8\left(1-r^2\right)$, and the volume of the region between them is 8π.

13. Use the same transformation as in the solution of Problem 10.

16. $M = \frac{6}{35}$; $M_y = \frac{55}{504} = M_x$; the centroid is at the point $\left(\frac{275}{432}, \frac{275}{432}\right) \approx (0.636574, 0.636574)$.

22. Note that $\bar{x} = \bar{y}$. The area of the quarter-ring is $A = \frac{1}{4}\left(\pi b^2 - \pi a^2\right) = \frac{1}{4}\pi\left(b^2 - a^2\right)$, and the volume generated by its rotation about the x-axis is $V = \frac{2}{3}\pi b^3 - \frac{2}{3}\pi a^3$. Therefore

$$V = \tfrac{2}{3}\pi\left(b^3 - a^3\right) = (2\pi\bar{y})\frac{\pi}{4}\left(b^2 - a^2\right).$$

a) Consequently $\bar{y} = \dfrac{(2/3)\,\pi\left(b^3 - a^3\right)}{(1/2)\,\pi^2\left(b^2 - a^2\right)} = \dfrac{4\left(b^2 + ab + a^2\right)}{3\pi\,(b+a)} = \bar{x}.$

b) $\displaystyle\lim_{b\to a} \bar{x} = \frac{12a^2}{(3\pi)(2a)} = \frac{2a}{\pi} = \lim_{b\to a} \bar{y}.$

25. $V = \displaystyle\int_0^{2\pi} \int_0^1 \left(r\sqrt{5-r^2} - 2r^2\right) dr\, d\theta.$

28. $M = \frac{1}{48}a^6$

34. $2\pi^2 a^2 b$

46. $8\left(\pi - 2\right)$

52. The Jacobian of the transformation is 2; the integral becomes $\displaystyle\int_0^1 \int_{-v}^{v} 2e^{u/v}\, du\, dv = e - \frac{1}{e} \approx 2.3504.$

55. b) Integration by parts:

$$\left[\begin{array}{ll} u = 8\sqrt{3}\arcsin\dfrac{1}{\sqrt{3-x^2}} & dv = dx \\[3mm] du = 8\sqrt{3}\dfrac{x}{(3-x^2)\sqrt{2-x^2}}\, dx & v = x \end{array}\right]$$

$$A = \left[8x\sqrt{3}\arcsin\frac{1}{\sqrt{3-x^2}}\right]_0^1 - 8\sqrt{3}\int_0^1 \frac{x^2}{(3-x^2)\sqrt{2-x^2}}\, dx$$

$$= 8\sqrt{3}\arcsin\frac{1}{\sqrt{2}} - 8\sqrt{3}\int_0^1 \frac{x^2}{(3-x^2)\sqrt{2-x^2}}\, dx$$

$$= 2\pi\sqrt{3} - 8\sqrt{3}\int_0^1 \frac{x^2}{(3-x^2)\sqrt{2-x^2}}\, dx.$$

Now let $x = \sqrt{2}\sin\theta$:

$$A = 2\pi\sqrt{3} - 8\sqrt{3}\int_0^{\pi/4} \frac{2\sin^2\theta}{(3-2\sin^2\theta)}\, d\theta$$

$$= 2\pi\sqrt{3} - 8\sqrt{3}\int_0^{\pi/4} \frac{1-\cos 2\theta}{3-2\sin^2\theta}\, d\theta = \frac{1}{2}\int \frac{1-\cos\phi}{2+\cos\phi}\, d\phi. \quad (\phi = 2\theta)$$

Finally substitute $u = \tan\left(\dfrac{\theta}{2}\right)$ to (eventually) obtain $A = 4\pi\left(\sqrt{3} - 1\right)$.

58. The average squared distance $\overline{d}$ of points of the ellipsoid from its center (at $(0,0,0)$) is

$$\frac{1}{V}\int_0^{2\pi}\int_0^{\pi}\int_0^1 (abc\rho^2\sin\phi)\rho^2\left[(a\sin\phi\cos\theta)^2 + (b\sin\phi\sin\theta)^2 + (c\cos\phi)^2\right]\,d\rho\,d\phi\,d\theta$$

where $V = \frac{4}{3}\pi abc$ is the volume of the ellipsoid. With the aid of a computer algebra system, we find that $\overline{d} = \frac{1}{5}(a^2 + b^2 + c^2)$. Compare this with the result of Problem 27 of Section 14.9; also examine the answer in the case that the ellipsoid is a sphere: $a = b = c$.

Chapter 15: Vector Calculus

Section 15.1

13. Matches Fig. 15.1.10

16. $\operatorname{div} \mathbf{F}(x, y, z) = 3 - 2 - 4 = -3;$ $\quad \operatorname{curl} \mathbf{F}(x, y, z) = \begin{vmatrix} \mathbf{i} & \mathbf{j} & \mathbf{k} \\ \dfrac{\partial}{\partial x} & \dfrac{\partial}{\partial y} & \dfrac{\partial}{\partial z} \\ 3x & -2y & -4z \end{vmatrix} = 0.$

22. $\operatorname{div} \mathbf{F}(x, y, z) = e^{xz} \cos y - e^{xy} \sin z;$ $\quad \operatorname{curl} \mathbf{F}(x, y, z) = \langle xe^{xy} \cos z - xe^{xz} \sin y, -ye^{xy} \cos z, ze^{xz} \sin y \rangle.$

34. $\nabla f \times \nabla g = \begin{vmatrix} \mathbf{i} & \mathbf{j} & \mathbf{k} \\ f_x & f_y & f_z \\ g_x & g_y & g_z \end{vmatrix} = \langle f_y g_z - f_z g_y, f_z g_x - f_x g_z, f_x g_y - f_y g_x \rangle.$

Therefore

$$\operatorname{div}(\nabla f \times \nabla g) = f_{yx} g_z + f_y g_{zx} - f_{zx} g_y - f_z g_{yz}$$
$$+ f_{zy} g_x + f_z g_{xy} - f_{xy} g_z - f_x g_{zy}$$
$$+ f_{xz} g_y + f_x g_{yz} - f_{yz} g_x - f_y g_{xz} = 0$$

wherever the mixed second-order partial derivatives of both f and g are equal: $f_{xy} = f_{yx}$, and so on.

40. $\dfrac{1}{r} = (x^2 + y^2 + z^2)^{-1/2}$, so $\nabla \left(\dfrac{1}{r} \right) = -\tfrac{1}{2} (x^2 + y^2 + z^2)^{-3/2} \langle 2x, 2y, 2z \rangle = -\dfrac{1}{r^3} \langle x, y, z \rangle = -\dfrac{1}{r^3} \mathbf{r}.$

41. $\operatorname{div}(r\mathbf{r}) = \operatorname{div}\left(\sqrt{x^2 + y^2 + z^2} \, \langle x, y, z \rangle \right)$

$$= \frac{\partial}{\partial x}\left(x\sqrt{x^2 + y^2 + z^2} \right) + \frac{\partial}{\partial y}\left(y\sqrt{x^2 + y^2 + z^2} \right) + \frac{\partial}{\partial z}\left(z\sqrt{x^2 + y^2 + z^2} \right)$$

$$= 3\sqrt{x^2 + y^2 + z^2} + \frac{x^2}{\sqrt{x^2 + y^2 + z^2}} + \frac{y^2}{\sqrt{x^2 + y^2 + z^2}} + \frac{z^2}{\sqrt{x^2 + y^2 + z^2}} = 3r + r = 4r.$$

Section 15.2

1. $dx = 4\,dt$, $dy = 3\,dt$, and $ds = 5\,dt$. Also $f(t) = (4t - 1)^2 + (3t + 1)^2 = 25t^2 - 2t + 2.$

Therefore $\displaystyle \int_C f(x, y)\,ds = \int_{-1}^{1} (25t^2 - 2t + 2)(5)\,dt = \frac{310}{3};$

$\displaystyle \int_C f(x, y)\,dx = \frac{248}{3}$ and $\displaystyle \int_C f(x, y)\,dy = 62.$

4. Note that $ds = dt$. So $\displaystyle \int_C (2x - y)\,ds = \int_0^{\pi/2} (2\sin t - \cos t)\,dt = 1;$

$$\int_C (2x - y)\,dx = \int_0^{\pi/2} \left(2\sin t \, \cos t - \frac{1 + \cos 2t}{2} \right) dt = \frac{4 - \pi}{4};$$

$$\int_C (2x - y)\,dy = \int_0^{\pi/2} (-2\sin^2 t + \sin t \, \cos t)\,dt = \frac{1 - \pi}{2}.$$

7. Use y as the parameter.

10. If $f(x, y) = \tfrac{1}{2}x^2 + 2xy - \tfrac{1}{2}y^2$, then $\nabla f = \langle x + 2y, 2x - y \rangle$. Therefore the given integral is independent of the path. Hence $\displaystyle \int_C (x + 2y)\,dx + (2x - y)\,dy = f(-2, -1) - f(3, 2) = -9.$

13. $\displaystyle\int_C \mathbf{F} \cdot \mathbf{T}\, ds = \int_{t=0}^{\pi} (y\, dx - x\, dy + z\, dz) = \int_0^{\pi} (\cos^2 t + \sin^2 t + 4t)\, dt = \pi + 2\pi^2.$

16. C is parallel to $\langle 2, 3, 3 \rangle$, so C may be parametrized by
$$\mathbf{r}(t) = \langle 1, -1, 2 \rangle + \langle 2t, 3t, 3t \rangle = \langle 2t+1, 3t-1, 3t+2 \rangle;$$
that is, we let $x = 2t+1$, $y = 3t-1$, and $z = 3t+2$ for $0 \le x \le 1$. Then $ds = \sqrt{22}\, dt$, so
$$\int_C xyz\, ds = \int_0^1 (2t+1)(3t-1)(3t+2)\sqrt{22}\, dt = 7\sqrt{22}.$$

19. If the wire has constant density k, then the total mass of the wire is $M = \pi a k$ and the mass of a segment of length ds is $dm = k\, ds$. Parametrize the wire as follows:
$$x = a\cos\theta, \quad y = a\sin\theta, \quad 0 \le \theta \le \pi.$$
Then compute $M_x = \displaystyle\int_C y\, dm.$

22. With the parametrization of the helical wire given in Problem 21, we find that $ds = 5\, dt$ and thus that $dm = 5k\, dt$. The mass of the wire is
$$m = \int_0^{2\pi} 5k\, dt = 10k\pi$$
and its moment of intertia around the z-axis is
$$I_z = \int_0^{2\pi} 5k[(x(t))^2 + (y(t))^2]\, dt = 90k\pi = 9m.$$

25. From the given parametrization we find that $ds = \frac{3}{2}|\sin 2t|\, dt$, and hence
$$I_0 = 4\int_0^{\pi/2} k(\cos^6 t + \sin^6 t)\, ds = \left[\frac{3}{64}k(-\cos 6t - 7\cos 2t) \right]_0^{\pi/2} = 3k.$$
Because the mass of the wire is
$$m = 4\int_0^{\pi/2} k\, ds = \left[-3k\cos 2t \right]_0^{\pi/2} = 6k,$$
we can also write $I_0 = m/2$.

28. We use the parametrization of the cycloid given in Problem 24. Then $ds = 2\sin(t/2)$, and so the cycloidal arch has length
$$\int_0^{2\pi} 2\sin\frac{t}{2}\, dt = 8.$$
Hence the average distance of points of the cycloidal arch from $(0,0)$ is
$$\overline{d} = \frac{1}{8}\int_0^{2\pi} (t^2 + 2 - 2\cos t - 2t\sin t)^{1/2}\, 2\sin\frac{t}{2}\, dt \approx 3.55252.$$

31. With the given parametrization, we find that $ds = \sqrt{2}\, e^{-t}\, dt$ and that
$$\sqrt{[x(t)]^2 + [y(t)]^2} = e^{-t}.$$
The length of the spiral is
$$\int_0^{\infty} \sqrt{2}\, e^{-t}\, dt = \left[-\sqrt{2}\, e^{-t} \right]_0^{\infty} = \sqrt{2},$$
so the average distance of points of the spiral from $(0,0)$ is
$$\overline{d} = \frac{1}{\sqrt{2}}\int_0^{\infty} \sqrt{2}\, e^{-2t}\, dt = \frac{1}{2}.$$

40. Parametrization: $x = 10t$, $y = t^2$, $10 \ge t \ge 0$.
$$W = \int_{10}^0 \langle 0, -150 \rangle \cdot \langle 10, 2t \rangle\, dt = \left[-150t^2 \right]_{10}^0 = 15,000 \quad \text{(ft-lb)}.$$

Section 15.3

1. If $\nabla f = \mathbf{F} = \langle 2x + 3y, 3x + 2y \rangle$ then $f_x(x, y) = 2x + 3y$, so $f(x, y) = x^2 + 3xy + g(y)$. Then $3x + g'(y) = f_y(x, y) = 3x + 2y$, so we may choose $g(y) = y^2$. Answer: One potential function for $\mathbf{F}$ is $f(x, y) = x^2 + 3xy + y^2$.

4. $f(x, y) = x^2 y^2 + x^3 + y^4$

10. $f(x, y) = \frac{1}{2}x^2 + x \arctan y + \frac{1}{2}\ln(1 + y^2)$

16. $f(x, y) = xy^{-2/3} + x^{-3/2}y = \dfrac{x^{5/2} + y^{5/3}}{y^{2/3}x^{3/2}}$

22. $\dfrac{\partial}{\partial y}(y^2 + 2xy) = 2y + 2x = \dfrac{\partial}{\partial x}(x^2 + 2xy)$, so the integral is path-independent in the entire plane. If $\mathbf{F}(x, y) = \langle y^2 + 2xy, x^2 + 2xy \rangle$, then $\mathbf{F}$ is the gradient of $f(x, y) = xy^2 + x^2 y$, so the value of the integral is $f(1, 2) - f(0, 0) = 4 + 2 = 6$.

28. $f(x, y, z) = x^2 - xy - xz + y^2 + z^2$

34. You should find that $f(x, y, z) = xyz + \frac{1}{2}y^2 + z$.

37. Substitution of the given data in the formula of Problem 36 (and remembering to convert kilometers into meters) yields $W \approx 8.04 \times 10^{10}$ N·m.

Section 15.4

1. $\displaystyle\int_{-1}^{1}\int_{-1}^{1}(2x - 2y)\,dy\,dx = \int_{-1}^{1} 4x\,dx = 0.$

4. $\displaystyle\int_{0}^{1}\int_{x^2}^{x}(y + 2y)\,dy\,dx = \frac{1}{5}.$

7. $\displaystyle\int_{0}^{\pi}\int_{0}^{\sin x} 1\,dy\,dx$

10. $Q_x - P_y = 0$, so the value of the integral is zero.

13. Let C denote the circle with the parametrization given in the problem. The area of the circular disk it bounds is given by

$$A = \oint_C x\,dy = \oint_C (a\cos t)(a\cos t)\,dt = a^2 \int_0^{2\pi} \frac{1}{2}(1 + \cos 2t)\,dt = \pi a^2.$$

16. The area is $\frac{5}{12}$.

19. $Q_x - P_y - 21x^2 y^2 - 15x^2 y^2 = 6x^2 y^2$, so the work is

$$W = \int_0^3 \int_0^{6-2x} 6x^2 y^2\,dy\,dx = \int_0^2 \left[2x^2 y^3\right]_0^{6-2x} dx$$

$$= \int_0^3 2x^2 (6 - 2x)^3\,dx = \left[144x^3 - 108x^4 + \tfrac{144}{5}x^5 - \tfrac{8}{3}x^6\right]_0^3 = \tfrac{972}{5}.$$

22. $\nabla \cdot \mathbf{F} = 3x^2 + 3y^2$, so

$$\phi = 4 \int_0^2 \int_0^{\sqrt{4-x^2}} (3x^2 + 3y^2)\, dy\, dx$$

$$= 4 \int_0^2 \left[3x^2 y + y^3 \right]_0^{\sqrt{4-x^2}} dx$$

$$= 4 \int_0^2 2(2 + x^2)\sqrt{4 - x^2}\, dx$$

$$= \left[2(2x + x^3)\sqrt{4 - x^2} + 48\arcsin\frac{x}{2} \right]_0^2 = 24\pi.$$

34. $A = \oint_C x\, dy = \int_0^\pi 2\sin t\, \cos^2 t\, dt = \frac{4}{3}.$

Section 15.5

1. $dS = \sqrt{3}\, dy\, dx$, so

$$\iint_S f(x, y, z)\, dS = \int_0^1 \int_0^{1-x} (x + y)\sqrt{3}\, dy\, dx = \frac{1}{3}\sqrt{3}.$$

4. $dS = r\sqrt{2}\, dr\, d\theta$, so

$$\iint_S f(x, y, z)\, dS = \int_0^{2\pi} \int_0^2 r^3\sqrt{2}\, dr\, d\theta = 8\pi\sqrt{2}.$$

7. $dS = r\sqrt{3}\, dr\, d\theta$, so

$$I_z = \iint_S \delta(x^2 + y^2)\, dS = \int_0^{2\pi} \int_0^3 r^3\sqrt{3}\, dr\, d\theta = \frac{81}{2}\pi\sqrt{3}.$$

10. $dS = r\sqrt{2}\, dr\, d\theta$, so

$$I_z = \iint_S \delta(x^2 + y^2)\, dS = \int_0^{2\pi} \int_2^5 r^3\sqrt{2}\, dr\, d\theta = \frac{609}{2}\pi\sqrt{2}.$$

13. Parametrize S as follows: $z = 3\cos\phi$, $0 \le \phi \le \frac{1}{2}\pi$, $0 \le \theta \le 2\pi$. An upper unit normal for S is $\mathbf{n} = \frac{1}{3}\langle x, y, z\rangle$, $dS = 9\sin\phi\, d\phi\, d\theta$, and $\mathbf{F}\cdot\mathbf{n} = 3\sin^2\phi$. Hence

$$\iint_S \mathbf{F}\cdot\mathbf{n}\, dS = \int_0^{2\pi} \int_0^{\pi/2} 27\sin^3\phi\, d\phi\, d\theta = 36\pi.$$

16. $\frac{16}{3}\pi$

19. On the bottom face of the cube (where $z = 0$, $0 \le x \le 1$, $0 \le y \le 1$), we take unit normal $\mathbf{n} = -\mathbf{k}$. But then $\mathbf{F}\cdot\mathbf{n} = -3z \equiv 0$ on the bottom face, so the flux across that surface is zero. It is also zero on the other two faces that lie in the coordinate planes. On the upper surface (where $z = 1$, $0 \le x \le 1$, $0 \le y \le 1$, we take unit normal $\mathbf{n} = \mathbf{k}$, and so $\mathbf{F}\cdot\mathbf{n} = 3z \equiv 3$. So the flux across that surface is $3 \cdot 1^2 = 3$. Similarly, the flux across the face $y = 1$, $0 \leq x \leq 1$, $0 \leq z \leq 1$ is 2 and the flux across the face $x = 1$, $0 \leq y \leq 1$, $0 \leq z \leq 1$ is 1. Therefore $\phi = 6$.

22. 32π

28. Assume unit density; $V = (2 - \sqrt{2})\, \pi a^2$; $M_{xy} = \frac{1}{2}\pi a^3$. Centroid: $\left(0, 0, \frac{1}{4}\left(2 + \sqrt{2}\right) a\right)$.

31. $dS = \sqrt{1 + 4y^2}$, so

$$I_z = \iint_S \delta(x^2 + y^2)\, dS = \int_{-1}^{1} \int_{-2}^{2} (x^2 + y^2)\sqrt{1 + 4y^2}\, dy\, dx = \frac{115}{12}\sqrt{17} + \frac{13}{48}\sinh^{-1} 4.$$

34. The formula is

$$\iint_S P\, dy\, dz + Q\, dz\, dx + R\, dx\, dy = \iint_D \left(P - Q\frac{\partial x}{\partial y} - R\frac{\partial x}{\partial z}\right) dy\, dz.$$

37. The given parametrization yields $\mathbf{N} = \langle -2bu^2 \cos v, -2au^2 \sin v, abu \rangle$, so the area of the paraboloid is

$$A = \int_0^{2\pi} \int_0^c u\sqrt{(2au \sin v)^2 + (2bu \cos v)^2 + (ab)^2}\, du\, dv$$

$$= \int_0^{2\pi} \frac{\left[(ab)^2 + 2(ac)^2 + 2(bc)^2 - 2(a^2 - b^2)c^2 \cos 2v\right]^{3/2} - (ab)^3}{6[a^2 + b^2 - (a^2 - b^2)\cos 2v]}\, dv.$$

We believe the last integral to be nonelementary (*Mathematica* 3.0 uses elliptic functions to compute its antiderivative), but with $a = 4$, $b = 3$, and $c = 2$ we find that its value is

$$\int_0^{2\pi} \frac{-1728 + (344 - 56 \cos 2v)^{3/2}}{6(25 - 7 \cos 2v)}\, dv \approx 194.703.$$

To find I_z, we insert the factor $\delta[(au \cos v)^2 + (bu \sin v)^2]$ into the previous integrand (with $\delta \equiv 1$); with $a = 4$, $b = 3$, and $c = 2$ a numerical integration produces the result $I_z \approx 5157.17$.

40. The given parametrization of the Möbius strip yields

$$|\mathbf{N}| = \sqrt{16 + \tfrac{3}{4}t^2 + 8t \cos\left(\tfrac{1}{2}\theta\right) + \tfrac{1}{2}t^2 \cos\theta},$$

and thus it has area

$$A = \int_0^{2\pi} \int_{-1}^{1} |\mathbf{N}|\, dt\, d\theta \approx 50.3986$$

and moment of inertia around the z-axis

$$I_z = \int_0^{2\pi} \int_{-1}^{1} (x^2 + y^2) \cdot |\mathbf{N}|\, dt\, d\theta \approx 831.47.$$

Section 15.6

1. First, div $\mathbf{F} = 3$. So $\iiint_B \text{div}\,\mathbf{F}\, dV = \iiint_B 3\, dV = (3)\left(\frac{4}{3}\right)(\pi)(1)^3 = 4\pi$.

A unit normal is $\mathbf{n} = \langle x, y, z \rangle$, and $\mathbf{F} \cdot \mathbf{n} = 1$. So $\iint_S \mathbf{F} \cdot \mathbf{n}\, dS = (1)(4\pi)(1)^2 = 4\pi$.

Both values are 4π; this verifies the divergence theorem in this special case.

4. Here, div $\mathbf{F} = y + z + x$; on the face where $z = 0$, $0 \leq x \leq 2$, and $0 \leq y \leq 2$, we have $\mathbf{F} \cdot \mathbf{n} = 0$. On the opposite face we find that $\mathbf{F} \cdot \mathbf{n} = 2x$. The surface integral of $\mathbf{F} \cdot \mathbf{n}$ on that face is equal to 8.

Similar results hold on the other four faces, and both integrals in the divergence theorem are equal to 24.

7. Let C denote the solid cylinder. Then

$$\iint_S \mathbf{F} \cdot \mathbf{n} \, dS = \iiint_C \left(3x^2 + 3y^2 + 3z^2\right) \, dV = 3 \int_0^{2\pi} \int_0^3 \int_{-1}^4 \left(r^2 + z^2\right) r \, dz \, dr \, d\theta.$$

10. Let B denote the solid bounded by S; note that div $\mathbf{F} = x^2 + y^2$. So

$$\iint_S \mathbf{F} \cdot \mathbf{n} \, dS = \iiint_B \operatorname{div} \mathbf{F} \, dV = \int_0^{2\pi} \int_0^3 \int_{r^2}^9 r^3 \, dz \, dr \, d\theta = \frac{243}{2}\pi.$$

Section 15.7

1. Because $\mathbf{n}$ is to be the upper unit normal, we have

$$\mathbf{n} = \mathbf{n}(x, y, z) = \tfrac{1}{2}\langle x, y, z \rangle.$$

The boundary curve C of the hemispherical surface S has the parametrization

$$x = 2\cos\theta, \quad y = 2\sin\theta, \quad 0 \le \theta \le 2\pi.$$

Therefore

$$\iint_S (\operatorname{curl} \mathbf{F}) \cdot \mathbf{n} \, dS = \int_C 3y \, dx - 2x \, dy + xyz \, dz = \int_0^{2\pi} \left(-12\sin^2\theta - 8\cos^2\theta\right) d\theta = -20\pi.$$

4. Parametrize the boundary curves as follows:

$$
\begin{aligned}
C_1: &\quad x = \cos t, \quad y = \sin t, \quad z = 1, \quad 0 \le t \le 2\pi; \\
C_2: &\quad x = \cos t, y = -\sin t, \quad z = 3, \quad 0 \le t \le 2\pi.
\end{aligned}
$$

Then

$$\iint_S (\operatorname{curl} \mathbf{F}) \cdot \mathbf{n} \, dS = \int_{C_1} \mathbf{F} \cdot \mathbf{T} \, ds + \int_{C_2} \mathbf{F} \cdot \mathbf{T} \, ds = \int_0^{2\pi} \left(2\sin^2 t - 2\cos^2 t\right) dt = 0.$$

7. Parametrize S (the elliptical region bounded by C) as follows:

$$x = z = r\cos t, \quad y = r\sin t, \quad 0 \le r \le 2, \quad 0 \le t \le 2\pi.$$

Then $\mathbf{r}_r \times \mathbf{r}_t = \langle -r, 0, r \rangle$, $dS = r\sqrt{2} \, dr \, dt$; the upper unit normal for S is

$$\mathbf{n} = \tfrac{1}{2}\sqrt{2}\langle -1, 0, 1 \rangle,$$

and $\operatorname{curl} \mathbf{F} = \langle 3, 2, 1 \rangle$. So $(\operatorname{curl} \mathbf{F}) \cdot \mathbf{n} = -\sqrt{2}$. Hence

$$\int_C \mathbf{F} \cdot \mathbf{T} \, ds = \iint_S (\operatorname{curl} \mathbf{F}) \cdot \mathbf{n} \, dS = \int_0^{2\pi} \int_0^2 (-2r) \, dr \, dt = -8\pi.$$

10. Let E be the ellipse bounded by C. Now E lies in the plane with equation $-y + z = 0$, so has upper unit normal

$$\mathbf{n} = \tfrac{1}{2}\sqrt{2}\langle 0, -1, 1 \rangle.$$

Next, curl $\mathbf{F} = \langle -2z, -2x, -2y \rangle$, so $(\text{curl } \mathbf{F}) \cdot \mathbf{n} = \sqrt{2}\,(x - y)$. The projection of E into the xy-plane may be described this way:

$$x^2 + (y - 1)^2 = 1, \quad z = 0;$$

alternatively, by $r = 2 \sin t$ $(0 \leq t \leq \pi)$, $z = 0$. Thus

$$\int_C \mathbf{F} \cdot \mathbf{T}\, ds = \iint_E (\text{curl } \mathbf{F}) \cdot \mathbf{n}\, dS = \iint_E \sqrt{2}\,(x - y)\, dS.$$

A parametrization of E is

$$\mathbf{w}\,(r, t) = \langle r \cos t, r \sin t, r \sin t \rangle, \quad 0 \leq t \leq \pi, \quad 0 \leq r \leq 2 \sin t.$$

Next we find that $\mathbf{n} = \mathbf{w}_r \times \mathbf{w}_t = \langle 0, -r, r \rangle$. Therefore $dS = r\sqrt{2}\, dr\, dt$. Consequently,

$$\iint_E \sqrt{2}\,(x - y)\, dS = \int_0^\pi \int_0^{2 \sin t} \sqrt{2}\,r\,(\cos t - \sin t)\, r\sqrt{2}\, dr\, dt.$$

Answer: -2π.

22. Another of the glories of Western civilization.

$$L = \iint_S (\mathbf{r} - \mathbf{r}_0) \times (-\rho g z \mathbf{n})\, dS = \rho g \iiint_V \text{curl}\,(z\,(\mathbf{r} - \mathbf{r}_0))\, dV \quad \text{(by Problem 20)}.$$

But $\text{curl}\,(z\,(\mathbf{r} - \mathbf{r}_0)) = (\nabla z) \times (\mathbf{r} - \mathbf{r}_0) + z\,(\text{curl}\,(\mathbf{r} - \mathbf{r}_0))$. It follows immediately that $\nabla z = \mathbf{k}$ and that $\text{curl}\,(\mathbf{r} - \mathbf{r}_0) = \mathbf{0}$. Thus

$$\mathbf{L} = \rho g \iiint_V \mathbf{k} \times (\mathbf{r} - \mathbf{r}_0)\, dV = \rho g \mathbf{k} \times \left(\iiint_V \mathbf{r}\, dV - \mathbf{r}_0 V \right).$$

consequently $\mathbf{L} = \mathbf{0}$ as desired, because $\mathbf{r}_0 = \dfrac{1}{V} \iiint_V \mathbf{r}\, dV$.

Chapter 15 Miscellaneous

1. C is part of the graph of $y = \frac{4}{3}x$ and $ds = \frac{5}{3}\, dx$. So $\displaystyle\int_C (x^2 + y^2)\, ds = \int_0^3 \left(x^2 + \left(\frac{4}{3}x \right)^2 \right) \frac{5}{3}\, dx = \frac{125}{3}$.

4. 115

7. If $\nabla \phi = \langle x^2 y, xy^2 \rangle$ then $\phi_x = x^2 y$, so that $\phi\,(x, y) = \frac{1}{3}x^3 y + g\,(y)$. Thus $\phi_y = \frac{1}{3}x^3 + g'\,(y) = xy^2$. This is impossible unless x is constant, but x is not constant on C.

10. Parametrize C: $x = t$, $y = t^2$, $z = t^3$, $1 \leq z \leq 2$.

$$W = \int_C P\, dx + Q\, dy + R\, dz = \int_C z\, dx - x\, dy + y\, dz = \int_1^2 (t^3 - 2t^2 + 3t^4)\, dt = \frac{1061}{60}.$$

16. Note that if $P_y < Q_x$ at some point of D, then $Q_x - P_y > 0$ on a small region R surrounding that point, so that

$$\iint_R (Q_x - P_y)\, dA > 0.$$

19. Checkpoints: The flux across the upper surface is 60π and the flux across the lower surface is 12π.

Appendices

Appendix A

1. $|3 - 17| = |-14| = 14$
4. $|5| - |-7| = 5 - 7 = -2$
7. $|(-3)^3| = |-27| = 27$
10. $-|7 - 4| = -|3| = -3$
13. $2x < 4$, so $x < 2$. Answer: $(-\infty, 2)$.
16. $2x \leq 4$, so $x \leq 2$. Answer: $(-\infty, 2]$.
19. $-8 < 2x < 2$, so $-4 < x < 1$. Answer: $(-4, 1)$.
22. $2 < -5x < 6$; $-6 < 5x < -2$; $-\frac{6}{5} < x < -\frac{2}{5}$. Answer: $\left(-\frac{6}{5}, -\frac{2}{5}\right)$.
25. We solve instead $|1 - 3x| \leq 2$: $-2 \leq 3x - 1 \leq 2$; $-1 \leq 3x \leq 3$; $-\frac{1}{3} \leq x \leq 1$. So the solution set of the original inequality is $\left(-\infty, -\frac{1}{3}\right) \cup (1, +\infty)$.
28. $1/(2x + 1) > 3$ implies that $2x + 1 > 0$, so $1 > 3 \cdot (2x + 1) = 6x + 3$; $-2 > 6x$; $x < -\frac{1}{3}$. But $2x + 1 > 0$, so $x > -\frac{1}{2}$. Answer: $\left(-\frac{1}{2}, -\frac{1}{3}\right)$.
31. If $x \neq \frac{1}{5}$, then $|1 - 5x| > 0$, so

$$\frac{1}{|1 - 5x|} > 0 > -\frac{1}{3}.$$

Answer: $\left(-\infty, \frac{1}{5}\right) \cup \left(\frac{1}{5}, +\infty\right)$.
43. If a and b are both negative, then $a < 0$, so $a \cdot b > 0 \cdot b$, and thus ab is positive. If a is positive and b is negative, then $b < 0$, so $a \cdot b < a \cdot 0 = 0$. Hence ab is negative.

Appendix B

1. Both the segments AB and BC have the same slope 1.
4. AB has slope 1 but BC has slope $\frac{8}{7}$.
7. AB has slope 2 and AC has slope $-\frac{1}{2}$.
10. $m = -1$, $b = 1$
13. $m = -\frac{2}{5}$, $b = \frac{3}{5}$
16. The points $(2, 0)$ and $(0, -3)$ lie on L, so the slope of L is $\frac{3}{2}$. An equation of L is $2y = 3x - 6$.
19. The slope of L is -1; an equation is $y - 2 = 4 - x$.
22. The *other* line has slope $-\frac{1}{2}$, so L has slope 2 and therefore equation $y - 4 = 2(x + 2)$.
25. The parallel lines have slope 5, so the line $y = -\frac{1}{5}x + 1$ is perpendicular to them both. It meets the line $y = 5x + 1$ at $(0, 1)$ and the line $y = 5x + 9$ at $\left(\frac{-20}{13}, \frac{17}{13}\right)$, and the distance between these two points is

$$\sqrt{\left(\frac{20}{13}\right)^2 + \left(\frac{4}{13}\right)^2} = \frac{4}{13}\sqrt{26} \approx 1.568929.$$

28. Show that AB is parallel to CD, that BC is parallel to AD, that AB and CD have the same length, that AD and BC have the same length, and that AC is perpendicular to BD by showing that the product of their slopes is -1.
31. The slope of $P_1 M$ is

$$\frac{y_1 - \bar{y}}{x_1 - \bar{x}} = \frac{y_1 - \frac{1}{2}(y_1 + y_2)}{x_1 - \frac{1}{2}(x_1 + x_2)} = \frac{2y_1 - y_1 - y_2}{2x_1 - x_1 - x_2} = \frac{y_1 - y_2}{x_1 - x_2},$$

and the slope of MP_2 is the same.

34. Begin with the fact that $L = mC + b$ for some constants m and b. The data given in Problem 34 yield

$$124.942 = 20m + b \text{ and } 125.134 = 110m + b.$$

It follows that $m = \frac{192}{90000}$ and $b = \frac{187349}{1500}$. Thus, approximately, $L = (0.00213)C + 124.899$.

37. $x = -2.75,\ y = 3.5$

40. $x = \frac{25}{4},\ y = \frac{3}{2}$

43. $x = -\frac{7}{4},\ y = \frac{33}{8}$

46. $x = \frac{59}{11},\ y = -\frac{12}{11}$

Appendix C

1. $40 \cdot \dfrac{\pi}{180} = \dfrac{2\pi}{9}$ (rad)

4. $210 \cdot \dfrac{\pi}{180} = \dfrac{7\pi}{6}$

7. $\dfrac{2\pi}{5} \cdot \dfrac{180}{\pi} = 72°$

10. $\dfrac{23\pi}{60} \cdot \dfrac{180}{\pi} = 69°$

	x	$\sin x$	$\cos x$	$\tan x$	$\cot x$	$\sec x$	$\csc x$
13.	$\frac{7}{6}\pi$	$-\frac{1}{2}$	$-\frac{1}{2}\sqrt{3}$	$\frac{1}{3}\sqrt{3}$	$\sqrt{3}$	$-\frac{2}{3}\sqrt{3}$	-2

16. $\sin x = 1$ for $x = \frac{1}{2}(4n + 1)\pi$ where n is an integer.

19. $\cos x = 1$ when x is an integral multiple of 2π.

22. $\tan x = 1$ for $x = \frac{1}{4}(4k + 1)\pi$ where k is an integer.

25. See the figure at the right.

$\sin x = -\frac{3}{5}$, $\cos x = \frac{4}{5}$, $\tan x = -\frac{3}{4}$,

$\csc x = -\frac{5}{3}$, $\sec x = \frac{5}{4}$, $\cot x = -\frac{4}{3}$.

31. $\sin(11\pi/6) = \sin(-\pi/6) = -\sin(\pi/6) = -1/2$

34. $\cos(4\pi/3) = -\cos(\pi/3) = -1/2$

37. **(a)** $\cos\left(\dfrac{\pi}{2} - \theta\right) = \cos\left(\dfrac{\pi}{2}\right)\cos\theta + \sin\left(\dfrac{\pi}{2}\right)\sin\theta = 0 \cdot \cos\theta + 1 \cdot \sin\theta = \sin\theta.$

40. $\tan(\pi \pm \theta) = \tan(\pm\theta)$ (because the tangent function is periodic with period π) and $\tan(\pm\theta) = \pm\tan(\theta)$ because the tangent function is odd (the quotient of an odd function and an even function).

43. $3\sin^2 x - 1 + \sin^2 x = 2$:
$4\sin^2 x = 3$;
$\sin x = \pm\frac{1}{2}\sqrt{3}$
$x = \frac{1}{3}\pi$, $x = \frac{2}{3}\pi$.

46. $2\sin^2 x + \cos x = 2$:
$2 - 2\cos^2 x + \cos x = 2$;
$(\cos x)(2\cos x - 1) = 0$;
$\cos x = 0$ or $\cos x = \frac{1}{2}$;
$x = \frac{1}{2}\pi$, $x = \frac{1}{3}\pi$.

Appendix D

1.   Given $\epsilon > 0$, choose $\delta = \epsilon$. If $0 < |x - a| < \delta$ then $|x - a| < \epsilon$. Therefore $\lim_{x \to a} x = a$.

4.   Given $\epsilon > 0$, choose $\delta = \epsilon/2$. If $0 < |x - (-3)| < \delta$ then $|2x + 6| < 2\delta$, so $|(2x + 1) - (-5)| < \epsilon$. Therefore, by definition, $\lim_{x \to -3} (2x + 1) = -5$.

7.   Given $\epsilon > 0$, let $\delta = \epsilon/k$ where you will choose a suitably large positive integer later in the proof so that the inequalities will work. Then go back through the proof and replace k by this integer. Write it neatly and turn it in for a perfect grade. This is a standard technique in ϵ-δ proofs.

10.   The desired inequality $\left| \dfrac{1}{\sqrt{x}} - \dfrac{1}{\sqrt{a}} \right| < \epsilon$ follows from $|\sqrt{a} - \sqrt{x}| < \epsilon\sqrt{ax}$, and we keep x from becoming negative or zero (or too close to zero) by requiring that $a/2 < x < 3a/2$. We ensure this by requiring that $\delta < a/2$. We obtain the inequality $|\sqrt{a} - \sqrt{x}| < \epsilon\sqrt{ax}$ from

$$\left| \sqrt{a} + \sqrt{x} \right| \left| \sqrt{a} - \sqrt{x} \right| < \epsilon\sqrt{ax} \left(\sqrt{a} + \sqrt{x} \right) ;$$

that is,

$$|x - a| < \epsilon\sqrt{ax} \left(\sqrt{a} + \sqrt{x} \right) .$$

It suffices that

$$\delta < \epsilon\sqrt{ax} \left(\sqrt{a} + \sqrt{x} \right) ,$$

and this can be assured by choosing $\delta < 3a^{3/2}\epsilon/4$, as this implies the last displayed inequality. So choose δ to be the minimum of $a/2$ and $3a^{3/2}\epsilon/4$.

13.   First deal with the case $L > 0$. The desired inequality follows from $|f(x) - L| < \epsilon L f(x)$. But $f(x)$ can be forced to lie between $L/2$ and $3L/2$ by choosing δ sufficiently small, and now the choice of δ is clear: Merely ensure that $\delta < L^2\epsilon/2$.

16.   Let $\epsilon = \frac{1}{2}f(a)$. Note that $f(x) \to f(a)$ as $x \to a$ by hypothesis. Now choose δ such that

$$|f(x) - f(a)| < \epsilon \quad \text{if} \quad 0 < |x - a| < \delta.$$

For such x, we have $-\epsilon < f(x) - 2\epsilon < \epsilon$, so $\epsilon < f(x) < 3\epsilon$. Therefore $f(x) > 0$ if x is in the interval $(a - \delta, a + \delta)$.

Appendix H

1. $\displaystyle\int_0^1 x^2 \, dx = \frac{1}{3}$.

4. $\displaystyle\int_0^3 \frac{x}{\sqrt{16 + x^2}} \, dx = \left[\sqrt{16 + x^2} \right]_0^3 = 1$.

7. $\displaystyle\int_0^2 \sqrt{x^4 + x^7} \, dx = \left[\frac{2}{9} \left(1 + x^3 \right)^{3/2} \right]_0^2 = \frac{52}{9}$.

Typeset by $\mathcal{A}\mathcal{M}\mathcal{S}$-TEX